计算机技术开发与应用丛书

Java多线程并发体系实战
微课视频版

刘宁萌 编著

清华大学出版社

北京

内容简介

本书全面、系统地介绍了与 Java 多线程相关的核心知识点,将官方多线程的核心知识点用链路的方式进行讲解。读者在学习的过程中需要循序渐进,核心知识点的理解是非常重要的,由核心知识点扩展开来就可以形成万物,希望读者在学习和创造的过程中能拥有自己的成长和思维。

全书共 10 章。第 1~3 章介绍了线程的核心知识点,第 4 章讲解了线程的局部变量,第 5 章介绍了 Lock 锁,第 6 章讲解了 atomic 原子包,第 7 章介绍了阻塞队列,第 8 章介绍了线程池,第 9 章讲解了线程同步器,第 10 章分析了 AQS 源码。书中每个章节都有相对应的代码验证环节,提供了大量应用实例,每章均附有习题。

本书适合有一定的 Java 基础知识,包括语法、结构、面向对象的编程概念,最好具有一定的并发编程基础的读者阅读,也可作为计算机相关专业学生的学习用书或参考教材,还可作为 Java 软件开发者的自学用书或研究人员的参考用书。

本书封面贴有清华大学出版社防伪标签,无标签者不得销售。
版权所有,侵权必究。举报: 010-62782989,beiqinquan@tup.tsinghua.edu.cn。

图书在版编目(CIP)数据

Java 多线程并发体系实战: 微课视频版/刘宁萌编著. —北京: 清华大学出版社,2023.9
(计算机技术开发与应用丛书)
ISBN 978-7-302-63792-9

Ⅰ. ①J… Ⅱ. ①刘… Ⅲ. ①JAVA 语言—程序设计 Ⅳ. ①TP312.8

中国国家版本馆 CIP 数据核字(2023)第 105802 号

责任编辑: 赵佳霓
封面设计: 吴 刚
责任校对: 申晓焕
责任印制: 刘海龙

出版发行: 清华大学出版社
网　　址: http://www.tup.com.cn, http://www.wqbook.com
地　　址: 北京清华大学学研大厦 A 座　　　邮　编: 100084
社 总 机: 010-83470000　　　　　　　　　　邮　购: 010-62786544
投稿与读者服务: 010-62776969, c-service@tup.tsinghua.edu.cn
质量反馈: 010-62772015, zhiliang@tup.tsinghua.edu.cn
课件下载: http://www.tup.com.cn, 010-83470236

印 装 者: 三河市天利华印刷装订有限公司
经　　销: 全国新华书店
开　　本: 186mm×240mm　　印　张: 19　　字　数: 430 千字
版　　次: 2023 年 9 月第 1 版　　　　　　　　印　次: 2023 年 9 月第 1 次印刷
印　　数: 1~2000
定　　价: 79.00 元

产品编号: 099423-01

前言
PREFACE

2005年左右民用的多核CPU正式进入市场，标志着一个新时代来临了。双核和多核处理器设计用于在一枚处理器中集成两个或多个完整执行内核，以支持同时执行并管理多个任务。以前的单核CPU相当于一个人工作，它有很多任务，并很快地在任务之间切换工作，给人们造成的感觉好像是同时在执行，例如我上了一个聊天软件，我又在打游戏，感觉好像在同时执行，实际上单核CPU只是很快速地在各个任务之间切换工作，而多核心CPU相当于有多个人在同时工作，这才是真正意义上的多核心同时工作，当然它们也会在各个任务之间切换调度。

Java的多线程Thread核心类早在JDK 1.0就已经有了，从一开始就确立了Java最基础的线程模型，并且这样的线程模型在后续的修补中，并未发生实质上的变更，可以说是一个具有传承性的良好设计。

随着多核CPU的到来，在2004年发布的JDK 1.5中加入了Executors线程池、Callable、Future异步任务支持、Lock锁、concurrent并发包下的大量工具类，标志着Java多线程的时代来临了。

本书特点

本书以JDK 17长期维护版本为示例，详细地介绍了与Java多线程并发相关的知识体系，让读者不仅知其然，而且知其所以然。本书对多线程相关的理论分门别类、层层递进地进行详细叙述和透彻分析，既体现了各知识点之间的联系，又兼顾了其渐进性。本书在介绍每个知识点时都给出了该知识点的应用场景，同时配合代码演示，让读者更能明白其实现原理、使用方式。本书丰富的免费配套资源包括源代码、笔记、完整的配套视频。

本书主要内容

第1章介绍了Thread线程对象的创建方式、启动方式、常用方法、优先级、守护线程。

第2章介绍了ThreadGroup线程组的概念、线程对象构造器、线程类常用方法、线程对象异常捕获、等待线程对象销毁、线程对象的优雅关闭。

第3章介绍了线程之间的协作，synchronized对象锁、线程死锁的产生、Object对象监视器、线程阻塞等待机制、线程唤醒机制、线程可见性、重排序、线程生命周期状态。

第4章介绍了ThreadLocal线程局部变量的使用、实现机制的核心概念、Reference弱引用。

第 5 章介绍了 Lock 锁接口的相关规范，以及 ReentrantLock、Condition、ReentrantReadWriteLock。

第 6 章介绍了 atomic 原子包、AtomicBoolean、AtomicInteger、AtomicReference、AtomicIntegerFieldUpdater、AtomicIntegerArray、LongAdder。

第 7 章介绍了 BlockingQueue 接口的相关规范，及其核心实现类 ArrayBlockingQueue、LinkedBlockingQueue、LinkedTransferQueue、SynchronousQueue、PriorityBlockingQueue、DelayQueue。

第 8 章介绍了 ThreadPoolExecutor 线程池整个链路的串联、ThreadPoolExecutor 线程池的使用，源码分析其核心概念、FutureTask 的核心概念、AbstractExecutorService 的使用、ScheduledThreadPoolExecutor 定时器的使用，源码分析其核心概念。

第 9 章介绍了线程同步器的使用、CountDownLatch、CyclicBarrier、Semaphore、Phaser。

第 10 章介绍了 AbstractQueuedSynchronizer 框架，详细分析了其源码，利用官方文档示例实现自己的同步锁、同步器。

读者对象

所有对 Java 多线程感兴趣的软件开发人员。

资源下载提示

素材（源码）等资源：扫描目录上方的二维码下载。

视频等资源：扫描封底的文泉云盘防盗码，再扫描书中相应章节中的二维码，可以在线学习。

致谢

特别感谢清华大学出版社的赵佳霓编辑。感谢她对本书专业且高效的审阅及对本书提出的建设性意见，有效推动了本书的出版。同时也感谢参与本书出版的所有人员，在大家的辛勤努力下，才有了本书的顺利出版。

刘宁萌

2023 年 2 月

目 录
CONTENTS

本书源码

第1章 多线程基础(▶ 54min) ·· 1

1.1 多线程带来的好处和问题 ·· 1
    1.1.1 多线程带来的好处 ·· 1
    1.1.2 多线程带来的问题 ·· 1

1.2 进程和线程 ·· 2

1.3 线程创建方式 ·· 3

1.4 线程启动方式 ·· 3

1.5 线程的概念及常用方法 ·· 6
    1.5.1 线程状态 ·· 6
    1.5.2 常用方法 ·· 7

1.6 线程的优先级 ·· 10

1.7 守护线程 ·· 11

小结 ·· 14

习题 ·· 14

第2章 多线程进阶(▶ 149min) ·· 16

2.1 ThreadGroup 类线程组 ·· 16
    2.1.1 构造器 ·· 16
    2.1.2 常用方法 ·· 18

2.2 Thread 线程对象构造器 ·· 25
    2.2.1 stackSize(栈大小) ·· 25
    2.2.2 共享线程局部变量 ·· 27

2.3 Thread 类常用方法 ·· 28

2.4 Thread 线程对象异常捕获 ·· 30

	2.4.1 线程对象自己的异常捕获器	30
	2.4.2 所属线程组对象异常捕获器	30
	2.4.3 Thread 类全局异常捕获器	32
2.5	等待线程对象销毁	32
2.6	线程对象优雅关闭	34
	2.6.1 中断相关方法	34
	2.6.2 官方响应中断的方法	37
小结		39
习题		39

第 3 章 多线程特性（▶ 132min） 41

3.1	引出 synchronized 对象锁	41
3.2	synchronized 对象锁	44
	3.2.1 标准对象	44
	3.2.2 class 对象	45
	3.2.3 锁特性	47
3.3	线程死锁的产生	49
	3.3.1 JConsole	50
	3.3.2 jstack	50
3.4	对象监视器	54
	3.4.1 wait()	54
	3.4.2 wait(long timeoutMillis)	55
	3.4.3 notify()	56
	3.4.4 notifyAll()	58
3.5	线程的可见性和重排序	61
	3.5.1 可见性	61
	3.5.2 重排序	62
3.6	线程生命周期状态	64
	3.6.1 NEW	64
	3.6.2 RUNNABLE	65
	3.6.3 BLOCKED	65
	3.6.4 WAITING	66
	3.6.5 TIMED_WAITING	67
	3.6.6 TERMINATED	68
小结		68
习题		69

第 4 章　ThreadLocal 线程局部变量（▶ 104min）……74

- 4.1　在方法链路中传递数据……74
- 4.2　引出线程局部变量……75
- 4.3　线程局部变量核心概念……76
  - 4.3.1　Thread 对象数据保存点……76
  - 4.3.2　线程局部变量操作入口……82
  - 4.3.3　线程局部变量数据操作功能……86
  - 4.3.4　弱引用介绍……93
- 小结……96
- 习题……96

第 5 章　Lock 锁（▶ 104min）……100

- 5.1　Lock 接口……100
- 5.2　ReentrantLock……101
  - 5.2.1　构造器……101
  - 5.2.2　常用方法……101
  - 5.2.3　公平锁或非公平锁……110
  - 5.2.4　自旋锁……111
- 5.3　Condition……112
- 5.4　ReentrantReadWriteLock……117
  - 5.4.1　构造器……117
  - 5.4.2　共享锁和互斥锁……117
  - 5.4.3　重入特性……120
  - 5.4.4　常用方法……122
- 小结……128
- 习题……128

第 6 章　atomic 原子包（▶ 117min）……135

- 6.1　AtomicBoolean……135
  - 6.1.1　构造器……135
  - 6.1.2　常用方法……136
- 6.2　AtomicInteger……141
  - 6.2.1　构造器……141
  - 6.2.2　常用方法……141
- 6.3　AtomicReference……145

6.3.1 构造器 …… 146
  6.3.2 常用方法 …… 146
 6.4 AtomicIntegerFieldUpdater …… 149
  6.4.1 对象创建 …… 149
  6.4.2 常用方法 …… 150
 6.5 AtomicIntegerArray …… 151
  6.5.1 构造器 …… 151
  6.5.2 常用方法 …… 151
 6.6 LongAdder …… 153
  6.6.1 实现方式 …… 153
  6.6.2 常用方法 …… 155
 小结 …… 155
 习题 …… 156

第7章 阻塞队列（▷ 216min） …… 158
 7.1 ArrayBlockingQueue …… 158
  7.1.1 构造器 …… 158
  7.1.2 常用方法 …… 158
 7.2 LinkedBlockingQueue …… 170
  7.2.1 构造器 …… 171
  7.2.2 常用方法 …… 171
 7.3 LinkedTransferQueue …… 183
  7.3.1 构造器 …… 184
  7.3.2 常用方法 …… 184
 7.4 SynchronousQueue …… 193
  7.4.1 构造器 …… 193
  7.4.2 常用方法 …… 195
 7.5 DelayQueue …… 203
  7.5.1 构造器 …… 203
  7.5.2 常用方法 …… 203
 7.6 PriorityBlockingQueue …… 212
  7.6.1 构造器 …… 212
  7.6.2 常用方法 …… 213
 小结 …… 219
 习题 …… 219

第8章 线程池（▷ 248min） …… 222
 8.1 ThreadPoolExecutor …… 224

	8.1.1 构造器 ·· 224
	8.1.2 常用方法 ······································· 225

8.2 FutureTask ·· 235

    8.2.1 构造器 ·· 235

    8.2.2 常用方法 ······································· 235

8.3 AbstractExecutorService ······························ 241

    8.3.1 构造器 ·· 241

    8.3.2 常用方法 ······································· 242

8.4 ScheduledThreadPoolExecutor ··················· 247

    8.4.1 构造器 ·· 247

    8.4.2 常用方法 ······································· 247

小结 ··· 252

习题 ··· 252

第 9 章 线程同步器（▶ 152min） 256

9.1 CountDownLatch ····································· 256

    9.1.1 构造器 ·· 256

    9.1.2 常用方法 ······································· 256

9.2 CyclicBarrier ·· 259

    9.2.1 构造器 ·· 259

    9.2.2 常用方法 ······································· 259

9.3 Semaphore ·· 262

    9.3.1 构造器 ·· 262

    9.3.2 常用方法 ······································· 263

9.4 Phaser ··· 268

    9.4.1 构造器 ·· 268

    9.4.2 常用方法 ······································· 268

小结 ··· 276

习题 ··· 276

第 10 章 AQS 源码分析 279

10.1 构造器 ·· 279

10.2 常用方法 ··· 280

10.3 ConditionObject ······································ 289

小结 ··· 291

第 1 章 多线程基础

本章详细介绍线程的基础知识、核心概念。多线程并不复杂,个人想说万物始于小、始于简单,一个简单的东西在发展中产生交叉后再生再交叉,如果站在最初的点去看这些衍生关系,则很简单,否则很复杂。

1.1 多线程带来的好处和问题

现在计算机已经全面进入了多核心的时代,民用计算机普遍是 4 核心、8 核心起步,多核心带来了更高效的性能,同步处理能力相较单核心有多倍提升。

1.1.1 多线程带来的好处

从硬件 CPU 层面来讲,多线程才能发挥多核心 CPU 的全部优势。

从软件层面来讲,单个线程也无法支撑现在的软件应用。打个现实中的比喻,有 1 家店是卖面包的,这个店只有 1 个人,假设营业员、面包师都是 1 个人在做,那么在只有 1 个客人的情况下,假设是可以完成任务的,但是当同时有 5 个客人或者 10 个客人时,在这种情况下很明显只有 1 个人的面包店是不可能同时支撑起这些客户的需求的。多线程在现代的软件应用中是必需的。

由于现在的框架很多封装得比较深,开发者可能在现实工作编码中并没有写过多线程的代码,那是由于框架底层已经做了封装,只需按规范开发使用就好,但是如果想要自研框架、自研中间件,或者去了解其他框架的底层原理,则多线程是一个绕不开的主题。

1.1.2 多线程带来的问题

本节讨论的多线程带来的问题是针对 Java 语言层面的,硬件和系统层面的问题不进行深入讨论。

1. 共同工作产生的问题

这里可以用现实中工作的例子做一个比喻,例如共有 1000 块砖,需要搬到 A 地点 100 块,需要搬到 B 地点 200 块。

如果只有一个工人做,则只需记住自己搬的数量和需要的总量就可以完成任务。

如果由多个工人同时做,则彼此就需要沟通协调才能完成任务。如果沟通协调出现问题就会导致任务失败,例如数量多了或者少了都是失败。这种现实中的例子实际还是比较简单的,在 Java 多线程的世界中,需要考虑更多的可能性,接下来会详细介绍。

2. 线程上下文切换带来的性能损耗

线程上下文切换是单核、多核 CPU 都会存在的问题。所谓上下文切换,就好比在做一个任务 A 时,突然接收到任务 B 的信息要马上处理,这时肯定需要把任务 A 相关的信息进行缓存,以便后续再回到任务 A 时能继续处理任务 A 的工作,这个就是上下文切换。由系统来调度 CPU 的执行,上下文切换过多会造成 CPU 性能浪费。目前偏标准的一个做法是创建的线程数量最好是 CPU 核心数量的 2 倍或 3 倍,这个是一个偏理想的值。

1) CPU 密集型程序

一个计算型的应用程序,CPU 使用率非常高,当多线程并发执行时,可以充分利用 CPU 所有的核心数量,例如 8 核心的 CPU,开 8 个线程执行时,在理想情况下此时的效率最高,有效减少 CPU 上下文切换所带来的性能损耗,因此对于 CPU 密集型的任务,线程数量在理想情况下等于 CPU 核心数量是最好的。

2) I/O 密集型程序

I/O 密集型程序是指以文件交互或者网络传输为主的程序,执行这种程序时 CPU 会有大量的时间处于等待任务的状态,这时可以适当提高线程数量,以保证 CPU 核心的最大利用率,例如 8 核心的 CPU,可以开 40 个线程或者更多的线程,可视情况而定。

Linux 无任务下的默认线程数量,如图 1-1 所示。

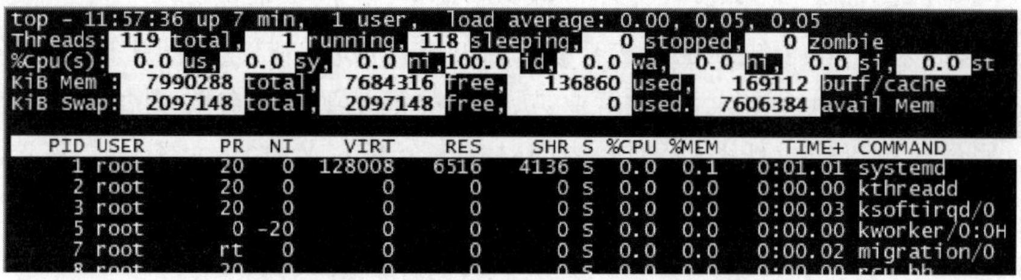

图 1-1 Linux 无任务下的默认线程数

1.2 进程和线程

进程是操作系统资源管理分配的基本单位,线程是操作系统任务调度和执行的基本单位。

在操作系统中能同时分配多个进程,而在同一个进程中又有多个线程,一个进程包含多个线程。程序在执行时,系统会为程序分配不同的内存空间,这个内存空间就可以理解为进程,而线程就在此内存空间执行代码调度。

用现实中的比喻来讲,就相当于公司或部门→人员的结构。这种抽象是无处不在的。

公司或部门只是一种层次上的管理抽象，具体的工作还是要由人员来做。这种现象有点类似进程→线程的关系。进程是系统抽象出来的一个管理空间的概念，程序在运行后，系统会给程序分配一个内存空间，这个内存空间称为进程。

1.3 线程创建方式

6min

Java 线程对象的创建方式，参考官方的 API 文档可知 Thread 类给出了两种方式。

1. 继承 Thread 类

一种是将类声明为 Thread 类的子类，此子类应重写超类的 run() 方法，代码如下：

```java
//第1章/three/ThreeThread.java
public class ThreeThread extends Thread{

    @Override
    public void run() {
        super.run();
    }
}
```

2. 实现 Runnable 接口

另一种方式是声明实现 Runnable 接口的子类。此子类应重写超类的 run() 方法，代码如下：

```java
//第1章/three/ThreeRunnable.java
public class ThreeRunnable implements Runnable{
    @Override
    public void run() {

    }
}
```

1.4 线程启动方式

12min

Java 线程的核心启动方式只有一种，别的启动方式都是基于核心启动方式的一种包装。

线程启动方式有且只有一种，也就是通过 Thread 对象的 start() 方法启动，代码如下：

```java
//第1章/four/FourThread.java
public class FourThread extends Thread{
```

```java
    @Override
    public void run() {
        try {
            Thread.sleep(10000);                    //睡眠 10s
        } catch (InterruptedException e) {
            throw new RuntimeException(e);
        }
        System.out.println("FourThread 执行了");
    }
}
//第 1 章/four/FourRunnable.java
public class FourRunnable implements Runnable{
    @Override
    public void run() {
        try {
            Thread.sleep(8000);                     //睡眠 8s
        } catch (InterruptedException e) {
            throw new RuntimeException(e);
        }
        System.out.println("FourRunnable 执行了");
    }
}
```

FourMain 类主方法,代码如下:

```java
//第 1 章/four/FourMain.java
public class FourMain {

    public static void main(String[] args) {
        System.out.println("start-" + System.currentTimeMillis());
            //这里既有默认的入口主线程 main,还有另外两个创建的线程启动执行
        FourThread p1 = new FourThread();
        p1.start();                          //启动线程

        FourRunnable p2 = new FourRunnable();
        new Thread(p2).start();              //启动线程
        System.out.println("end-" + System.currentTimeMillis());
    }
}
```

执行结果如下:

```
start-1659944267623
end-1659944267630
FourRunnable 执行了
FourThread 执行了
```

注意:这里一定要理解线程启动后的概念。

主方法是 Java 虚拟机（JVM）的入口，主方法运行后会有 3 个线程启动。一个是默认的主线程，还有两个是通过线程对象.start()方法启动的线程，如图 1-2 所示。

```
public static void main(String[] args) {
    System.out.println("start-" + System.currentTimeMillis());
    FourThread p1 = new FourThread();
    p1.start();        //启动线程

    FourRunable p2 = new FourRunable();
    new Thread(p2).start();        //启动线程
    System.out.println("end-" + System.currentTimeMillis());
}
```

图 1-2　线程启动

Thread 类 start()方法的源代码删减版，只保留了核心的内容，如图 1-3 所示。线程对象 start()方法执行后，最终会调用底层的 start0()方法，start0()方法是由 native 修饰的，是 Java 虚拟机底层的方法，start0()方法执行后，最终会启动线程并回调此线程对象的 run() 方法。

线程对象的两种创建方式造就了 run()方法执行后得到两种可能性。一种是继承自 Thread 类的，相当于直接重写了 run()方法，另外一种就是实现了 Runnable 接口的，最终通过回调 target.run()方法，回调到 Runnable 接口实现类的 run()方法。

不管使用线程创建方式中的哪一种，最终都会回调指定对象的 run()方法，所以在前面实现类时都要重写 run()方法，如图 1-4 所示。

```
public synchronized void start() {
    boolean started = false;
    try {
        start0();
        started = true;
    } finally {

    }
}
private native void start0();
```

图 1-3　Thread 类 start()核心源代码

```
@Override
public void run() {
    if (target != null) {
        target.run();
    }
}
```

图 1-4　官方 Thread 类 run()方法

修改 FourMain 类，代码如下：

```java
//第 1 章/four/FourMain.java
public class FourMain {

//单个线程执行,即主线程 main
    public static void main(String[] args) {
        System.out.println("start");
        FourThread p1 = new FourThread();
        p1.run();
```

```
            FourRunnable p2 = new FourRunnable();
            new Thread(p2).run();
            System.out.println("end");
        }
    }
```

执行结果如下：

```
start
FourThread 执行了
FourRunnable 执行了
end
```

注意：观察执行结果，并思考线程启动的方式。

1.5 线程的概念及常用方法

在 Java 的世界中万物皆对象，线程也被抽象成一种对象。那么在这里就有两个概念，一个是线程本身是 CPU 调度可以执行的代码，当然在这里指的是 Java 代码，另一个是 Java 设计概念，把线程抽象成了对象。

JVM 允许应用程序同时运行多个执行线程。每个线程都有一个优先级，具有较高优先级的线程优于较低优先级的线程执行。每个线程可能会也可能不会被标记为守护线程。当在某个执行线程中创建一个新的 Thread 线程对象时，新线程对象的优先级最初被设置为等于当前执行线程对象的优先级，并且当且仅当执行线程对象是守护线程时，新线程对象才是守护线程。当 JVM 启动时，通常有一个非守护线程（通常运行某个指定类的名为 main (String[] args) 的方法，称为主方法，默认是一个非守护线程），JVM 继续运行，直到所有非守护线程都已经退出或销毁，JVM 关闭。

1.5.1 线程状态

Thread.State 是一个枚举类。一个线程在给定的时间点只能处于一种状态。这些状态是不反映任何操作系统线程状态的 JVM 线程状态。

1. NEW

尚未启动的线程处于此状态。

2. RUNNABLE

在 JVM 中执行的线程处于此状态。

3. BLOCKED

阻塞等待拿锁的线程处于此状态。

4. WAITING

无限期等待的线程处于此状态。

5. TIMED_WAITING

最大等待指定时间的线程处于此状态。

6. TERMINATED

已退出的线程处于此状态。

1.5.2 常用方法

1. 线程名称

1) 通过构造参数传入名称

代码如下：

```java
//第1章/five/FiveThread.java
public class FiveThread extends Thread{
    public FiveThread() {
    }
    //构造器
    public FiveThread(String name) {
        super(name);
    }

    @Override
    public void run() {
        System.out.println("FiveThread:" + this.getName());         //当前对象名称
        System.out.println("FiveThread:" + Thread.currentThread().getName());
                                                                    //当前执行线程对象名称
    }
}
```

FiveMain 类主方法，代码如下：

```java
//第1章/five/FiveMain.java
public class FiveMain {

    public static void main(String[] args) {
        FiveThread fiveThread = new FiveThread("通过构造方法-名称");
        fiveThread.start(); //启动线程

    }
}
```

执行结果如下：

```
FiveThread:通过构造方法-名称
FiveThread:通过构造方法-名称
```

2）通过 setName(String name)方法设置名称

代码如下：

```java
//第 1 章/five/FiveMain.java
//include<five/FiveThread.java>
public class FiveMain {

    public static void main(String[] args) {
        FiveThread fiveThread = new FiveThread("通过构造方法-名称");
        fiveThread.setName("通过方法设置-名称");
        fiveThread.start(); //启动线程
    }
}
```

执行结果如下：

```
FiveThread:通过方法设置-名称
FiveThread:通过方法设置-名称
```

注意：如果没有设置名称，则默认生成一个固定格式的名称，如图 1-5 所示。

```
public Thread() {
    this( group: null,  target: null,  name: "Thread-" + nextThreadNum(),  stackSize: 0);
}
```
int值每次加1，并发安全

图 1-5　默认线程名称

线程的名称需要根据业务设置，做到见名知意，名称虽然没有强制要求唯一，但是在使用中不要出现同名的情况。

2. 线程 ID

getId()方法可以返回一个 long 型的值，每次创建线程对象时增加 1 操作，并发安全，如图 1-6 所示。

```java
private static long threadSeqNumber;

private static synchronized long nextThreadID() {
    return ++threadSeqNumber;
}
```

图 1-6　线程 ID

3. 当前执行线程对象

Thread.currentThread()可以获得当前执行线程对象，这里一定要分清楚当前对象和当前执行线程对象的区别，代码如下：

```java
//第 1 章/five/FiveThread.java
public class FiveThread extends Thread{
```

```java
    public FiveThread() {
    }
    //构造器
    public FiveThread(String name) {
        super(name);
    }

    @Override
    public void run() {
        System.out.println("FiveThread:" + this.getName());        //当前对象名称
        System.out.println("currentThread:" +
                Thread.currentThread().getName());
                                                                   //当前执行线程对象名称
    }
}
```

FiveMain 类主方法,代码如下:

```java
//第1章/five/FiveMain.java
public class FiveMain {

    public static void main(String[] args) {
        FiveThread fiveThread = new FiveThread("通过构造方法-名称");
        fiveThread.setName("通过方法设置-名称");
        fiveThread.start(); //启动线程
    }
}
```

执行结果如下:

```
FiveThread:通过方法设置-名称
currentThread:通过方法设置-名称
```

修改 FiveMain 类,代码如下:

```java
//第1章/five/FiveMain.java
public class FiveMain {

    public static void main(String[] args) {
        FiveThread fiveThread = new FiveThread("通过构造方法-名称");
        fiveThread.setName("通过方法设置-名称");
        fiveThread.run(); //修改此方法
    }
}
```

执行结果如下:

```
FiveThread:通过方法设置-名称
currentThread:main
```

> **注意**：fiveThread.run()方法并没有启动新的线程，里面是由当前主方法的默认线程去执行的，所以 Thread.currentThread().getName()输出的当前执行线程对象的名称是 main，而 this.getName()获得的是当前对象的名称。

1.6 线程的优先级

5min

创建新线程对象的优先级，默认从当前执行线程对象中获得并初始化，官方源代码的缩减版本只保留核心关键的代码，如图 1-7 所示。

```
private Thread(ThreadGroup g, Runnable target, String name,
               long stackSize, AccessControlContext acc,
               boolean inheritThreadLocals) {
    //源代码缩减版本
    this.name = name;
    Thread parent = currentThread();    // 当前执行线程对象
    SecurityManager security = System.getSecurityManager();
    g.addUnstarted();
    this.group = g;
    this.daemon = parent.isDaemon();
    this.priority = parent.getPriority();    // 初始化优先级
    if (security == null || isCCLOverridden(parent.getClass()))
        this.contextClassLoader = parent.getContextClassLoader();
    else
        this.contextClassLoader = parent.contextClassLoader;
    this.inheritedAccessControlContext =
            acc != null ? acc : AccessController.getContext();
    this.target = target;
    setPriority(priority);    // 底层方法，设置优先级
    if (inheritThreadLocals && parent.inheritableThreadLocals != null)
        this.inheritableThreadLocals =
            ThreadLocal.createInheritedMap(parent.inheritableThreadLocals);
    this.stackSize = stackSize;
    this.tid = nextThreadID();
}
```

图 1-7 Thread 类构造器中默认的优先级

官方默认提供了 3 个优先级的常量，数值范围是 1～10，优先级是一个相对的概念，并不是一个绝对的概念。优先级越高被执行的概率越大，如图 1-8 所示。

```
static final int    MAX_PRIORITY     线程可以具有的最大优先级
static final int    MIN_PRIORITY     线程可以具有的最低优先级
static final int    NORM_PRIORITY    分配给线程的默认优先级
```

图 1-8 Thread 类优先级常量

1. 获得优先级

可以通过 Thread 线程对象的 getPriority()方法获得优先级，代码如下：

```java
//第1章/six/SixThread.java
public class SixThread extends Thread{
    public SixThread() {
    }
    //构造器
    public SixThread(String name) {
        super(name);
    }

    @Override
    public void run() {
        System.out.println("level:" + this.getPriority());
    }
}
```

2. 设置优先级

可以通过 Thread 线程对象的 setPriority(int newPriority)方法设置优先级，代码如下：

```java
//第1章/six/SixMain.java
public class SixMain {

    public static void main(String[] args) {
        System.out.println("main:" + Thread.currentThread().getPriority());
        SixThread sixThread = new SixThread("通过构造方法-名称");
        sixThread.setPriority(Thread.MAX_PRIORITY);
        sixThread.start(); //启动线程

    }
}
```

执行结果如下：

```
main:5
level:10
```

1.7 守护线程

当JVM启动时，通常有一个非守护线程（通常指运行某个指定类的名为main的主方法）称为主线程。JVM继续运行，直到发生以下情况：

（1）已调用Runtime对象的退出方法，并且安全管理器已允许进行退出操作。
（2）所有不是守护线程的线程都已经被销毁。

1. 获得当前线程对象是否为守护线程

可以通过 Thread 线程对象的 isDaemon()方法获得此线程是否为守护线程。如果此线

12min

程是守护线程,则返回值为 true,否则返回值为 false。

2. 设置当前线程对象是否为守护线程

可以通过 Thread 线程对象的 setDaemon(boolean on)方法将此线程对象标记为守护线程或非守护线程,当所有执行的线程都是守护线程时,JVM 退出。

在默认情况下创建线程对象时,会从当前执行线程对象中获得并初始化,如图 1-9 所示。

```
private Thread(ThreadGroup g, Runnable target, String name,
               long stackSize, AccessControlContext acc,
               boolean inheritThreadLocals) {
    //源代码缩减版本
    this.name = name;
    Thread parent = currentThread();          ← 当前执行线程对象
    SecurityManager security = System.getSecurityManager();
    g.addUnstarted();
    this.group = g;
    this.daemon = parent.isDaemon();           ← 初始化是否守护线程
    this.priority = parent.getPriority();
    if (security == null || isCCLOverridden(parent.getClass()))
        this.contextClassLoader = parent.getContextClassLoader();
    else
        this.contextClassLoader = parent.contextClassLoader;
    this.inheritedAccessControlContext =
            acc != null ? acc : AccessController.getContext();
    this.target = target;
    setPriority(priority);
    if (inheritThreadLocals && parent.inheritableThreadLocals != null)
        this.inheritableThreadLocals =
            ThreadLocal.createInheritedMap(parent.inheritableThreadLocals);
    this.stackSize = stackSize;
    this.tid = nextThreadID();
}
```

图 1-9 初始化守护线程状态

当平时使用 main(String[] args)主方法运行时,主方法的执行线程是非守护线程,所以正常情况下使用的是非守护线程。当 JVM 里没有非守护线程时,JVM 退出,代码如下:

```java
//第 1 章/seven/SevenThread.java
public class SevenThread extends Thread{

    @Override
    public void run() {
        try {
            Thread.sleep(5000);                    //睡眠 5s
        } catch (InterruptedException e) {
            throw new RuntimeException(e);
        }
        System.out.println("SevenThread:isDaemon = " +
                            Thread.currentThread().isDaemon());
```

```
        }                               //是否为守护线程
    }
}
```

SevenMain 类主方法，代码如下：

```
//第1章/seven/SevenMain.java
public class SevenMain{

    public static void main(String[] args) {
        System.out.println(Thread.currentThread().getName());    //main
        System.out.println(Thread.currentThread().isDaemon());   //false

        SevenThread sevenThread = new SevenThread();
        sevenThread.setDaemon(true);                             //设置为守护线程
        sevenThread.start();

    }
}
```

执行结果如下：

```
main
false
```

观察结果，可以看到并没有输出 SevenThread 线程对象中的内容，因为它已被设置为守护线程，需要等待 5s 才会输出内容，但是在此过程中唯一的非守护线程 main 很快执行完毕了，所以 SevenThread 线程对象还在睡眠的过程中，此时的 JVM 环境里全是守护线程，JVM 退出，导致 SevenThread 线程对象直接被关闭了。

修改 SevenMain 类，代码如下：

```
//第1章/seven/SevenMain.java
public class SevenMain {

    public static void main(String[] args) {
        System.out.println(Thread.currentThread().getName());    //main
        System.out.println(Thread.currentThread().isDaemon());   //false

        SevenThread sevenThread = new SevenThread();
        sevenThread.start();                                     //启动线程

    }
}
```

执行结果如下：

```
main
false
SevenThread:isDaemon = false
```

可以看到输出了 SevenThread 线程对象中的内容,因为此时它是非守护线程,JVM 必须等待它执行完毕。

小结

本章介绍了线程的创建方式、启动方式、常用的几个概念方法,一定要理解线程启动后的概念,多线程执行和单线程执行的程序存在本质上的区别。当前对象和当前执行线程对象的区别也是一个很重要的概念。

注意: 在修改线程优先级或者判断是否为守护线程时,一定要在线程 start() 启动之前设置才会有效果。同一个线程对象只能使用一次 start() 方法启动线程。

习题

1. 判断题

(1) 线程的启动方式是线程对象的 run() 方法。(   )
(2) 线程的启动方式是线程对象的 start() 方法。(   )
(3) 设置线程的优先级,必须在线程没有启动前才有效果。(   )
(4) 当前对象和当前执行线程对象,可能相同也可能不同。(   )
(5) 默认执行 main 主方法的线程是非守护线程。(   )
(6) 当所有线程都是守护线程时,JVM 退出。(   )
(7) main 主方法执行完毕后,JVM 退出,不管有没有其他的线程。(   )
(8) 优先级 10 的线程比优先级 5 的线程,执行的概率绝对高 2 倍。(   )
(9) 优先级越高被优先执行的概率越高。(   )

2. 选择题

(1) 可以通过 Thread 线程对象调用的方法有(   )。(多选)
    A. start()　　　　　　B. run()　　　　　　C. start0()　　　　　　D. getPriority()
(2) 获得当前执行线程对象正确的方法是(   )。(单选)
    A. this　　　　　　　　　　　　　　　B. Thread.currentThread()
    C. this.getName()　　　　　　　　　　D. Thread.currentThread().getName()

3. 填空题

以下代码执行后的输出结果是_____。
代码如下:

```
//第1章/answer/TestThread.java
public class TestThread extends Thread{
```

```java
    public TestThread(TestRunnable testRunnable) {
        super(testRunnable);
    }

    @Override
    public void run() {
        System.out.println("TestThread.执行了");
    }
}

public class TestRunnable implements Runnable{

    @Override
    public void run() {
        System.out.println("Runnable.执行了");
    }
}
```

TestMain 类主方法,代码如下:

```java
public class TestMain {

    public static void main(String[] args) {
        TestRunnable testRunnable = new TestRunnable();
        TestThread testThread = new TestThread(testRunnable);
        testThread.start(); //启动线程
    }
}
```

第 2 章 多线程进阶

本章详细介绍 ThreadGroup 类线程组的概念、Thread 线程对象构造器、Thread 类常用方法、Thread 线程对象异常捕获、等待线程对象销毁、线程对象优雅关闭。

▷ 16min

2.1 ThreadGroup 类线程组

一个 ThreadGroup 线程组对象代表一组 Thread 线程对象集合,此外一个线程组对象还可以包含其他线程组对象,可以理解为当前线程组对象的下级线程组对象集合,线程组对象之间形成一个链路,每个线程组对象只能有一个上级线程组对象,但是可以有多个下级线程组对象集合,除了 Java 虚拟机初始线程组对象之外的每个线程组对象都有一个上级线程组对象。

▷ 2min

ThreadGroup 类核心字段说明,如图 2-1 所示。

```
public class ThreadGroup implements Thread.UncaughtExceptionHandler {
    private final ThreadGroup parent;  ← 上级线程组对象
    String name;
    int maxPriority;
    boolean destroyed;
    boolean daemon;

    int nUnstartedThreads = 0;  ← 没有启动的线程对象数量
    int nthreads;               ← 已经启动的线程对象数量
    Thread threads[];           ← 线程对象集合

    int ngroups;                ← 下级线程组对象数量
    ThreadGroup groups[];       ← 下级线程组对象集合
```

图 2-1 ThreadGroup 类核心字段

▷ 4min

▷ 14min

2.1.1 构造器

ThreadGroup 构造器见表 2-1。

第2章 多线程进阶

表 2-1 ThreadGroup 构造器

构 造 器	描 述
ThreadGroup(String name)	构造新的线程组对象,指定名称
ThreadGroup(ThreadGroup parent,String name)	构造新的线程组对象,指定上级线程组对象,指定名称

▶ 14min

1. ThreadGroup(String name)

接收 String 入参,作为当前线程组对象的名称,默认从当前执行线程对象中获得并初始化上级线程组对象,如图 2-2 所示。

▶ 11min

```
public ThreadGroup(String name) {      ← 构造器入口
    this(Thread.currentThread().getThreadGroup(), name);
}                                      ← 当前执行线程对象
public ThreadGroup(ThreadGroup parent, String name) {
    this(checkParentAccess(parent), parent, name);
}
private ThreadGroup(Void unused, ThreadGroup parent, String name) {
    this.name = name;
    this.maxPriority = parent.maxPriority;
    this.daemon = parent.daemon;
    this.parent = parent;              ← 初始化上级线程组对象
    parent.add(this);
}
```

图 2-2 构造器

parent.add(this)把当前线程组对象添加到上级线程组对象数组中,如图 2-3 所示。

```
private final void add(ThreadGroup g){
    synchronized (this) {
        if (destroyed) {
            throw new IllegalThreadStateException();
        }
        if (groups == null) {
            groups = new ThreadGroup[4];      ← 初始化数组
        } else if (ngroups == groups.length) {
            groups = Arrays.copyOf(groups, ngroups * 2);
        }                                     ← 扩容
        groups[ngroups] = g;                  ← 存放下级线程组对象

        ngroups++;   ← 索引增加,线程组对象数量
    }
}
```

图 2-3 存放下级线程组对象

2. ThreadGroup(ThreadGroup parent,String name)

接收 ThreadGroup 入参,作为当前线程组对象的上级线程组对象;接收 String 入参,作为当前线程组对象的名称,如图 2-4 所示。

```
public ThreadGroup(ThreadGroup parent, String name) {    ← 构造器入口
    this(checkParentAccess(parent), parent, name);
}                                                         ← 指定使用传入的parent
private ThreadGroup(Void unused, ThreadGroup parent, String name) {
    this.name = name;
    this.maxPriority = parent.maxPriority;
    this.daemon = parent.daemon;
    this.parent = parent;
    parent.add(this);
}
```

图 2-4　指定上级线程组对象

2.1.2　常用方法

1. getParent()

获得上级线程组对象，返回此线程组对象的上级线程组对象，代码如下：

```
//第2章/one/OneMain.java
public class OneMain {
    public static void main(String[] args) {
        ThreadGroup threadGroup1 = new ThreadGroup("my-1");
                //创建 threadGroup1 对象，指定名称
        ThreadGroup threadGroup2 = new ThreadGroup(threadGroup1,"my-2");
                //创建 threadGroup2 对象，指定上级线程组对象、名称
        ThreadGroup parent = threadGroup2.getParent();
        while (parent != null){
                     //循环获得上级线程组对象，直到其为空
            System.out.println(parent);
            parent = parent.getParent();
        }
    }
}
```

执行结果如下：

```
java.lang.ThreadGroup[name=my-1,maxpri=10]
java.lang.ThreadGroup[name=main,maxpri=10]
java.lang.ThreadGroup[name=system,maxpri=10]
```

注意：观察上方的输出内容，并理解除了 JVM 初始线程组对象之外的，每个线程组对象都有一个上级线程组对象。

2. activeCount()

获得此线程组内活动线程对象数量的估计值，返回 int 值，当创建 Thread 线程对象时，如果没有指定 ThreadGroup 线程组对象，则此线程对象归属于当前执行线程对象的线程组，如图 2-5 所示。

```
private Thread(ThreadGroup g, Runnable target, String name,
               long stackSize, AccessControlContext acc,
               boolean inheritThreadLocals) {   ← 构造器
    if (name == null) {
        throw new NullPointerException("name cannot be null");
    }

    this.name = name;

    Thread parent = currentThread();
    SecurityManager security = System.getSecurityManager();
    if (g == null) {   ← 如果为空则初始化
        if (security != null) {
            g = security.getThreadGroup();
        }
        if (g == null) {
            g = parent.getThreadGroup();
        }
    }
}
```

图 2-5 当线程组对象为空时初始化

当创建 Thread 线程对象时,也可以使用构造器指定归属于某个线程组对象,如图 2-6 所示。

```
public Thread( @Nullable ThreadGroup group, @NotNull String name) {
    this(group, target: null, name, stackSize: 0);
}
```

图 2-6 指定归属于某个线程组对象

当创建 Thread 线程对象时,会调用所属线程组对象的 addUnstarted() 方法,此方法会把线程组对象中 nUnstartedThreads 字段增加 1 操作,此字段代表当前线程组对象中所包含的未启动线程对象的数量,如图 2-7 所示。

```
private Thread(ThreadGroup g, Runnable target, String name,
               long stackSize, AccessControlContext acc,
               boolean inheritThreadLocals) {   ← 源代码精简版本
    if (name == null) {
        throw new NullPointerException("name cannot be null");
    }
    this.name = name;
    Thread parent = currentThread();
    SecurityManager security = System.getSecurityManager();
    if (g == null) {

        if (security != null) {
            g = security.getThreadGroup();
        }

        if (g == null) {
            g = parent.getThreadGroup();
        }
    }
    g.addUnstarted();   ← 调用线程组对象方法
}
```

图 2-7 线程构造器精简版

Thread 线程对象 start()启动时,会调用所属线程组对象的 add()方法,如图 2-8 所示。

```
public synchronized void start() {
    if (threadStatus != 0)
        throw new IllegalThreadStateException();
    group.add(this);  ← 线程对象中的group线程组对象
    boolean started = false;
    try {
        start0();
        started = true;
    } finally {
        try {
            if (!started) {
                group.threadStartFailed(this);
            }
        } catch (Throwable ignore) {

        }
    }
}
```

图 2-8　线程对象中的 group 字段

线程组对象的 add(Thread t)方法,把指定线程对象存放到线程组对象的 threads[]数组字段中,并更新 nthreads 字段(增加 1 操作),此字段代表当前线程组对象所包含的已启动线程对象的数量,然后更新 nUnstartedThreads 字段(减 1 操作),如图 2-9 所示。

```
void add(Thread t) {
    synchronized (this) {
        if (destroyed) {
            throw new IllegalThreadStateException();
        }
        if (threads == null) {
            threads = new Thread[4];  ← 初始化
        } else if (nthreads == threads.length) {
            threads = Arrays.copyOf(threads, nthreads * 2);
        }
        threads[nthreads] = t;  ← 存放线程对象
        nthreads++;  ← 已启动线程对象数量更新
        nUnstartedThreads--;  ← 未启动线程对象数量更新
    }
}
```

图 2-9　线程组对象 add(Thread t)方法

activeCount()返回此线程组对象及其后代组对象中已启动线程对象数量的估计值,可能有新加入的线程对象,也可能有结束生命周期并从组内删除的线程对象,所以这里说的是估计值,如图 2-10 所示。

```java
public int activeCount() {
    int result;

    int ngroupsSnapshot;
    ThreadGroup[] groupsSnapshot;
    synchronized (this) {
        if (destroyed) {
            return 0;
        }
        result = nthreads;
        ngroupsSnapshot = ngroups;
        if (groups != null) {
            groupsSnapshot = Arrays.copyOf(groups, ngroupsSnapshot);
        } else {
            groupsSnapshot = null;
        }
    }
    for (int i = 0 ; i < ngroupsSnapshot ; i++) {
        result += groupsSnapshot[i].activeCount();
    }
    return result;
}
```

图 2-10　线程对象数量的估计值

代码如下：

```java
//第2章/one/OneMain.java
public class OneMain {
    public static void main(String[] args) {
        ThreadGroup threadGroup1 = new ThreadGroup("my-1");
                //创建 threadGroup1 对象,指定名称
        ThreadGroup threadGroup2 = new ThreadGroup(threadGroup1,"my-2");
                //创建 threadGroup2 对象,指定上级线程组对象、名称
        new Thread(threadGroup2,"老唐").start();
            //创建线程对象,指定其归属的线程组对象、名称,然后启动线程
        System.out.println(threadGroup2.activeCount());
    }
}
```

执行结果如下：

```
1
```

修改 OneMain 类,代码如下：

```java
//第2章/one/OneMain.java
public class OneMain {
    public static void main(String[] args) throws InterruptedException {

        ThreadGroup threadGroup1 = new ThreadGroup("my-1");
        ThreadGroup threadGroup2 = new ThreadGroup(threadGroup1,"my-2");
        new Thread(threadGroup2,"老唐").start();
            //创建线程对象,指定其归属的线程组对象、名称,然后启动线程
        Thread.sleep(3000);                //睡眠 3s
        System.out.println(threadGroup2.activeCount());
    }
}
```

执行结果如下：

```
0
```

注意：观察上方的输出内容，并理解线程生命周期结束后，会从所属线程组对象内删除此线程对象。

Thread 线程对象生命周期结束前，JVM 会回调线程对象的 exit()方法，如图 2-11 所示。

```java
private void exit() {
    if (threadLocals != null && TerminatingThreadLocal.REGISTRY.isPresent()) {
        TerminatingThreadLocal.threadTerminated();
    }
    if (group != null) {
        group.threadTerminated(this);
        group = null;
    }

    target = null;

    threadLocals = null;
    inheritableThreadLocals = null;
    inheritedAccessControlContext = null;
    blocker = null;
    uncaughtExceptionHandler = null;
}
```

图 2-11 Thread 线程对象 exit()方法

3. activeGroupCount()

获得线程组对象数量估计值，返回此线程组对象及其后代组对象数量的估计值，代码如下：

```java
//第 2 章/one/OneMain.java
public class OneMain {
    public static void main(String[] args) throws InterruptedException {

        ThreadGroup threadGroup1 = new ThreadGroup("my-1");
                    //创建 threadGroup1 对象,指定名称
        ThreadGroup threadGroup2 = new ThreadGroup(threadGroup1,"my-2");
                    //创建 threadGroup2 对象,指定上级线程组对象、名称
        System.out.println(threadGroup1.getParent());
        System.out.println(threadGroup1.getParent().activeGroupCount());
    }
}
```

执行结果如下：

```
java.lang.ThreadGroup[name=main,maxpri=10]
2
```

4. enumerate(Thread[] list, boolean recurse)

将活动线程对象复制到指定的数组中,接收 Thread[] 入参,作为存放线程对象的数组容器,接收 boolean 入参,作为是否需要递归的条件,代码如下:

```java
//第 2 章/one/OneMain.java
public class OneMain {
    public static void main(String[] args) throws InterruptedException {

        ThreadGroup threadGroup1 = new ThreadGroup("my-1");
                    //创建 threadGroup2 对象,指定名称
        ThreadGroup threadGroup2 = new ThreadGroup(threadGroup1,"my-2");
                //创建 threadGroup2 对象,指定上级线程组对象、名称
        new Thread(threadGroup2, "老唐").start();
                //创建线程对象,指定其归属的线程组对象、名称,然后启动线程
        Thread[] threads = new Thread[5];
        threadGroup1.enumerate(threads,true);
        System.out.println(Arrays.toString(threads));

    }
}
```

执行结果如下:

[Thread[老唐,5,my-2], null, null, null, null]

修改 OneMain 类,代码如下:

```java
//第 2 章/one/OneMain.java
public class OneMain {
    public static void main(String[] args) throws InterruptedException {

        ThreadGroup threadGroup1 = new ThreadGroup("my-1");
        ThreadGroup threadGroup2 = new ThreadGroup(threadGroup1,"my-2");

        new Thread(threadGroup2, "老唐").start();
        Thread[] threads = new Thread[5];
        threadGroup1.enumerate(threads,false);

        System.out.println(Arrays.toString(threads));

    }
}
```

执行结果如下:

[null, null, null, null, null]

5. getMaxPriority()

返回此线程组对象的最大优先级。

6. getName()

返回此线程组对象的名称。

7. setMaxPriority(int pri)

设置此线程组及其子组的最大优先级，接收 int 入参，作为优先级，代码如下：

```java
//第2章/one/OneMain.java
public class OneMain {
    public static void main(String[] args) throws InterruptedException {

        ThreadGroup threadGroup1 = new ThreadGroup("my-1");
        ThreadGroup threadGroup2 = new ThreadGroup(threadGroup1,"my-2");
        threadGroup2.setMaxPriority(2); //设置最大优先级
        Thread thread = new Thread(threadGroup2, "老唐");
        thread.start();
        System.out.println(thread.getPriority());

        System.out.println("threadGroup2:" + threadGroup2.getMaxPriority());
    }
}
```

执行结果如下：

```
2
threadGroup2:2
```

当创建 Thread 线程对象时，会从所属线程组对象获得优先级，并更新线程对象的优先级，如图 2-12 所示。

```
private Thread(ThreadGroup g, Runnable target, String name,
               long stackSize, AccessControlContext acc,
               boolean inheritThreadLocals) {
    if (name == null) {
        throw new NullPointerException("name cannot be null");
    }
    this.name = name;
    Thread parent = currentThread();
    SecurityManager security = System.getSecurityManager();

    setPriority(priority);    ← 调用方法
}
public final void setPriority(int newPriority) {
    ThreadGroup g;
    checkAccess();
    if (newPriority > MAX_PRIORITY || newPriority < MIN_PRIORITY) {
        throw new IllegalArgumentException();
    }
    if((g = getThreadGroup()) != null) {
        if (newPriority > g.getMaxPriority()) {
            newPriority = g.getMaxPriority();    ← 从所属线程组对象更新优先级
        }
        setPriority0(priority = newPriority);
    }
}
```

图 2-12　从所属线程组对象更新优先级

2.2 Thread 线程对象构造器

Thread 构造器见表 2-2。

表 2-2 线程对象构造器

构 造 器	描 述
Thread(Runnable target)	构造新的线程对象,指定 Runnable 实现类
Thread(Runnable target,String name)	构造新的线程对象,指定 Runnable 实现类,指定名称
Thread(String name)	构造新的线程对象,指定名称
Thread(ThreadGroup group, Runnable target)	构造新的线程对象,指定所属线程组对象,指定 Runnable 实现类
Thread(ThreadGroup group, Runnable target,String name)	构造新的线程对象,指定所属线程组对象,指定 Runnable 实现类,指定名称
Thread(ThreadGroup group, Runnable target,String name,long stackSize)	构造新的线程对象,指定所属线程组对象,指定 Runnable 实现类,指定名称,指定堆栈大小
Thread(ThreadGroup group, Runnable target, String name, long stackSize, boolean inheritThreadLocals)	构造新的线程对象,指定所属线程组对象,指定 Runnable 实现类,指定名称,指定堆栈大小,指定是否共享线程局部变量
Thread(ThreadGroup group,String name)	构造新的线程对象,指定所属线程组对象,指定名称

以上线程对象构造器中的大部分已经使用过了,其实就是不同的入参组合形式,原理是一样的。这里主要应关注两个入参 stackSize、inheritThreadLocals。

2.2.1 stackSize(栈大小)

JVM 栈大小是用来管理方法区链路的,一般情况下使用默认值,如图 2-13 所示。

图 2-13 方法区链路

特殊情况下，当方法区链路过长时会抛出 StackOverflowError 异常，代码如下：

```java
//第 2 章/two/TwoMain.java
public class TwoMain {

    public static void main(String[] args) {
        TwoMain.recursion(100000);
    }

    //递归调用方法
    public static void recursion(int nums){
        if (nums != 0) {
            recursion( -- nums);
        }
    }
}
```

执行结果如下：

```
Exception in thread "main" java.lang.StackOverflowError
    at cn.kungreat.book.two.two.TwoMain.recursion(TwoMain.java:17)
```

1. 可以使用 -Xss 增加 JVM 栈大小

使用 IDEA 开发工具设置，如图 2-14 所示。

图 2-14　JVM 选项

IDEA 主方法运行语法，如图 2-15 所示。

```
"C:\Program Files\Java\jdk-17.0.3.1\bin\java.exe" -Xss128m
"-javaagent:C:\JetBrains\lib\idea_rt.jar=59521
        :C:\JetBrains\bin"
-Dfile.encoding=UTF-8
-classpath C:\Users\kungreat\IdeaProjects\bookteach\out\production\bookteach
        cn.kungreat.book.two.two.TwoMain
```

图 2-15　IDEA 主方法运行语法

2．构造线程对象时指定 stackSize

默认构造器 stackSize 的值为 0，此时由 JVM 管理分配栈大小。也可以自定义数值，当使用自定义数值时，只会影响此线程对象，代码如下：

```java
//第 2 章/two/TwoMain.java
public class TwoMain {

    public static void main(String[] args) {
        new Thread(Thread.currentThread().getThreadGroup(),() ->{
            TwoMain.recursion(100000);
            System.out.println(Thread.currentThread().getName());
        },"kkkk",99999999).start();
        //创建线程对象,指定其归属的线程组对象、Runnable实现类、
        //名称、栈大小,然后启动线程
    }

    //递归调用方法
    public static void recursion(int nums){
        if (nums != 0) {
            recursion(--nums);
        }
    }
}
```

执行结果如下：

```
kkkk
```

注意：将上方 TwoMain.java 代码中的 stackSize 修改为 10 000，并查看结果。

2.2.2　共享线程局部变量

inheritThreadLocals 此入参决定了是否共享当前执行线程对象的 inheritableThreadLocals 变量，如图 2-16 所示。

```
private Thread(ThreadGroup g, Runnable target, String name,
               long stackSize, AccessControlContext acc,
               boolean inheritThreadLocals) {      ← 源代码缩减版本

    this.name = name;

    if (inheritThreadLocals && parent.inheritableThreadLocals != null)
        this.inheritableThreadLocals =
            ThreadLocal.createInheritedMap(parent.inheritableThreadLocals);
    this.stackSize = stackSize;                    ← 共享
    this.tid = nextThreadID();
}
```

图 2-16　共享当前执行线程对象变量

2.3　Thread 类常用方法

1. activeCount()

静态方法，用于获得当前执行线程对象的组内活动线程对象数量的估计值，返回 int 值，如图 2-17 所示。

```
public static int activeCount() {
    return currentThread().getThreadGroup().activeCount();
}
```

图 2-17　当前执行线程对象的组内活动线程对象数量的估计值

2. enumerate(Thread[] tarray)

静态方法，用于将当前执行线程对象的组内活动线程对象复制到指定的数组中，接收 Thread[] 入参，作为存放线程对象的数组容器，如图 2-18 所示。

```
public static int enumerate(Thread tarray[]) {
    return currentThread().getThreadGroup().enumerate(tarray);
}
```

图 2-18　将当前执行线程对象的组内活动线程对象复制到指定的数组中

3. getAllStackTraces()

静态方法，用于返回所有活动线程对象的堆栈跟踪映射，代码如下：

```java
//第 2 章/three/ThreeMain.java
public class ThreeMain {

    public static void main(String[] args) throws InterruptedException {

        Map<Thread, StackTraceElement[]> allStackTraces = Thread.getAllStackTraces();
        System.out.println(allStackTraces);

    }
}
```

执行结果如下：

```
{Thread[main,5,main] = [Ljava.lang.StackTraceElement;@214c265e,
Thread[Signal Dispatcher,9,system] = [Ljava.lang.StackTraceElement;@448139f0,
Thread[Common-Cleaner,8,InnocuousThreadGroup] = [Ljava.lang.StackTraceElement;@7cca494b,
Thread[Notification Thread,9,system] = [Ljava.lang.StackTraceElement;@7ba4f24f, Thread
[Monitor Ctrl-Break,5,main] = [Ljava.lang.StackTraceElement;@3b9a45b3,
Thread[Reference Handler,10,system] = [Ljava.lang.StackTraceElement;@7699a589,
Thread[Finalizer,8,system] = [Ljava.lang.StackTraceElement;@58372a00,
Thread[Attach Listener,5,system] = [Ljava.lang.StackTraceElement;@4dd8dc3]}
```

观察 Debug 模式下的线程对象堆栈跟踪映射，如图 2-19 所示。

图 2-19　堆栈跟踪映射

4. onSpinWait()

静态方法，并且是一个空的实现，如图 2-20 所示。

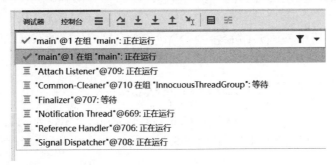

图 2-20　onSpinWait 源代码

即使没有调用 onSpinWait 方法，上面的代码也将保持正确，然而在某些架构上，JVM 可能会发出处理器指令，以更有益的方式处理此类代码逻辑，如图 2-21 所示。

```
class EventHandler {
    volatile boolean eventNotificationNotReceived;
    void waitForEventAndHandleIt() {
        while (eventNotificationNotReceived) {
            java.lang.Thread.onSpinWait();
        }
        readAndProcessEvent();
    }
    void readAndProcessEvent() {

    }
}
```

图 2-21　官方文档示例代码

5. sleep(long millis)

静态方法，为当前执行线程对象休眠（暂时停止执行）指定毫秒数，具体取决于系统计时器和调度程序的精度和准确性，接收 long 入参，作为最大等待时间的毫秒数。

6. sleep(long millis,int nanos)

静态方法,为当前执行线程对象休眠(暂时停止执行)指定毫秒数加上指定的纳秒数,具体取决于系统计时器和调度程序的精度和准确性,接收 long 入参,作为最大等待时间的毫秒数,接收 int 入参,作为指定的最大等待时间的纳秒数。

7. yield()

静态方法,向调度程序提示此线程愿意放弃其当前对 CPU 的使用,降低被处理器执行的概率。

2.4 Thread 线程对象异常捕获

线程对象异常捕获共有 3 个阶段,第 1 个阶段是自己的异常捕获器,第 2 个阶段是所属组的异常捕获器,第 3 个阶段是全局的异常捕获器。

2.4.1 线程对象自己的异常捕获器

通过 setUncaughtExceptionHandler(Thread.UncaughtExceptionHandler eh)方法设置此线程对象的异常捕获器,当此线程由于未捕获的异常而突然终止时调用的处理程序,代码如下:

```java
//第 2 章/four/FourMain.java
//include < four/FourThread.java >
public class FourMain {

    public static void main(String[] args) throws InterruptedException {
        FourThread fourThread = new FourThread();
            //创建 fourThread 对象,并设置对象的异常捕获器,匿名内部类实现形式
        fourThread.setUncaughtExceptionHandler(
            new Thread.UncaughtExceptionHandler(){
            @Override
            public void uncaughtException(Thread t, Throwable e) {
                System.out.println("第 1 个阶段,进入了对象的异常处理:" +
                    e.getMessage());
            }
        });
        fourThread.start(); //启动线程
    }
}
```

执行结果如下:

第 1 个阶段,进入了对象的异常处理:/ by zero

2.4.2 所属线程组对象异常捕获器

线程组对象默认的异常处理会一直迭代调用上级线程组对象的异常处理,直到上级线

程组对象为空,最终会调用 Thread 类全局的异常处理器,如图 2-22 所示。

```
public void uncaughtException(Thread t, Throwable e) {
    if (parent != null) {
        parent.uncaughtException(t, e);
    } else {
        Thread.UncaughtExceptionHandler ueh =
            Thread.getDefaultUncaughtExceptionHandler();
        if (ueh != null) {
            ueh.uncaughtException(t, e);
        } else if (!(e instanceof ThreadDeath)) {
            System.err.print("Exception in thread \""
                             + t.getName() + "\" ");
            e.printStackTrace(System.err);
        }
    }
}
```

图 2-22 线程组对象默认异常处理

可以重写线程组对象默认异常处理,代码如下:

```
//第 2 章/four/FourGroup.java
public class FourGroup extends ThreadGroup{
    public FourGroup(String name) {
        super(name);
    }
    //重写线程组对象的默认异常处理方法
    @Override
    public void uncaughtException(Thread t, Throwable e) {
        System.out.println("第 2 个阶段,进入了对象的异常处理:" + e.getMessage());
    }
}
```

FourMain 类主方法,代码如下:

```
//第 2 章/four/FourMain.java
public class FourMain {

    public static void main(String[] args) throws InterruptedException {

        FourGroup fourGroup = new FourGroup("第 2 个阶段");
        FourThread fourThread = new FourThread(fourGroup,"kungreat");
        fourThread.start();
    }
}
```

执行结果如下:

```
第 2 个阶段,进入了对象的异常处理:/ by zero
```

2.4.3 Thread 类全局异常捕获器

通过 setDefaultUncaughtExceptionHandler(Thread.UncaughtExceptionHandler eh) 方法设置全局异常捕获器，当线程对象由于未捕获的异常而突然终止时调用的默认处理程序，代码如下：

```java
//第 2 章/four/FourMain.java
public class FourMain {

    public static void main(String[] args) throws InterruptedException {
        Thread.setDefaultUncaughtExceptionHandler(
            new Thread.UncaughtExceptionHandler(){
            @Override
            public void uncaughtException(Thread t, Throwable e) {
                System.out.println("第 3 个阶段,进入了对象的异常处理:"
                                                    + e.getMessage());
            }
        });                          //设置全局异常捕获器,匿名内部类实现形式
        FourThread fourThread = new FourThread();
        fourThread.start();          //启动线程

    }
}
```

执行结果如下：

第 3 个阶段,进入了对象的异常处理:/ by zero

2.5 等待线程对象销毁

8min

在多线程并发执行的场景下,有些需求存在依赖关系。例如 A、B 两个任务同时执行,但是在某个阶段 B 任务必须依赖于 A 任务的完成(A 线程销毁)才能继续执行,join() 方法提供了此种实现,代码如下：

```java
//第 2 章/five/FiveMain.java
public class FiveMain {

    public static void main(String[] args) throws InterruptedException {
        Thread thread = new Thread(new Runnable() {
            @Override
            public void run() {
                try {
                    Thread.sleep(5000);          //睡眠 5s
                    System.out.println("A 任务开始...");
                } catch (InterruptedException e) {
                    throw new RuntimeException(e);
```

```
                    }
                    System.out.println("A 任务结束...");
                }
            });
            thread.start();                            //启动线程

            System.out.println("B 任务开始...");
            thread.join();                             //等待线程对象销毁
            System.out.println("B 任务结束...");
        }

}
```

执行结果如下：

```
B 任务开始...
A 任务开始...
A 任务结束...
B 任务结束...
```

join(long millis)等待此线程对象销毁，最大等待毫秒数，接收 long 入参，作为最大等待时间的毫秒数。

join()方法源代码依赖于 wait(0)方法，一直等待直到线程对象销毁，如图 2-23 所示。

```
public final void join() throws InterruptedException {
    join(0);
}

public final synchronized void join(final long millis)
throws InterruptedException {
    if (millis > 0) {
        if (isAlive()) {
            final long startTime = System.nanoTime();
            long delay = millis;
            do {
                wait(delay);
            } while (isAlive() && (delay = millis -
                    TimeUnit.NANOSECONDS.toMillis(System.nanoTime() - startTime)) > 0);
        }
    } else if (millis == 0) {
        while (isAlive()) {         ┌─ 一直等待线程对象死亡
            wait(0);
        }
    } else {
        throw new IllegalArgumentException("超时值为负");
    }
}
```

图 2-23　join()方法

2.6 线程对象优雅关闭

此功能依赖于interrupt()中断此线程对象的相关方法,此方法是一种设计模式的体现,在线程执行的过程中是不可预期的,所以不能强制销毁正在执行的线程。在编码过程中,需要根据业务去处理响应中断标记,这里还要特别注意,官方也提供了响应中断标记的方法,也要考虑此因素可能带来的影响。

2.6.1 中断相关方法

1. isInterrupted()

测试此线程对象是否已中断,返回boolean值,代码如下:

```java
//第 2 章/six/SixMain.java
public class SixMain {

    public static void main(String[] args) throws InterruptedException {
        Thread thread = new Thread(new Runnable() {
            @Override
            public void run() {
                System.out.println("one:" +
                    Thread.currentThread().isInterrupted());
                                    //测试此线程对象是否已中断
                System.out.println("one:" +
                    Thread.currentThread().isInterrupted());
            }
        },"one");
        thread.start(); //启动线程

        System.out.println("main:" +
            Thread.currentThread().isInterrupted());
        System.out.println("main:" +
            Thread.currentThread().isInterrupted());
    }
}
```

执行结果如下:

```
main:false
main:false
one:false
one:false
```

2. interrupt()

中断此线程对象,此功能类似一个标记,需要编码处理中断响应逻辑,代码如下:

```java
//第 2 章/six/SixMain.java
public class SixMain {
```

```java
    public static void main(String[] args) throws InterruptedException {
        Thread thread = new Thread(new Runnable() {
            @Override
            public void run() {
                int num = 0;
                while (true){
                    Thread.yield();                    //当前线程愿意放弃当前对处理器的使用
                    System.out.println("...执行了第" + num + "次任务");
                    num++;
                    if(Thread.currentThread().isInterrupted()){
                        System.out.println("中断此任务,保存必要的数据");
                        return;
                    }
                }
            }
        },"one");
        thread.start();                       //启动线程

        Thread.sleep(1000);                   //睡眠1s
        thread.interrupt();
    }
}
```

执行结果如下：

```
...执行了第 312399 次任务
...执行了第 312400 次任务
...执行了第 312401 次任务
中断此任务,保存必要的数据
```

修改 SixMain 类,代码如下：

```java
//第2章/six/SixMain.java
public class SixMain {

    public static void main(String[] args) throws InterruptedException {
        Thread thread = new Thread(new Runnable() {
            @Override
            public void run() {
                int num = 0;
                while (true){

                    Thread.yield();                    //当前线程愿意放弃当前对处理器的使用
                    System.out.println("...执行了第" + num + "次任务");
                    num++;
                }
            }
        },"one");
        thread.start();
```

```
            Thread.sleep(1000);
            thread.interrupt();
        }
    }
```

注意：执行上方代码，并理解中断的设计概念。

3. interrupted()

静态方法，用于测试当前执行线程对象是否已中断，有清除中断的效果，返回 boolean 值，代码如下：

```
//第2章/six/SixMain.java
public class SixMain {

    public static void main(String[] args) throws InterruptedException {
        Thread thread = new Thread(new Runnable() {
            @Override
            public void run() {
                System.out.println("one:" + Thread.interrupted());
                System.out.println("one:" + Thread.interrupted());
            }
        },"one");
        thread.interrupt();                    //中断此线程
        thread.start();                        //启动线程
        System.out.println("main:" +
            Thread.currentThread().isInterrupted());
                                    //测试此线程是否已中断
        System.out.println("main:" +
            Thread.currentThread().isInterrupted());
    }
}
```

执行结果如下：

```
main:false
main:false
one:true
one:false
```

此方法的官方源代码，如图 2-24 所示。

```
public static boolean interrupted() {
    Thread t = currentThread();
    boolean interrupted = t.interrupted;
    if (interrupted) {
        t.interrupted = false;
        clearInterruptEvent();
    }
    return interrupted;
}
```

图 2-24 interrupted()方法源代码

2.6.2 官方响应中断的方法

当线程对象有中断标记时,如果该线程对象在执行中调用 Object 对象的 wait()、wait(long timeoutMillis)、wait(long timeoutMillis, int nanos)方法,或者 Thread 线程对象的 join()、join(long millis)、join(long millis, int nanos)方法,或者 Thread 类的 sleep(long millis)、sleep(long millis, int nanos)方法,则其中断状态将被清除并抛出 InterruptedException 异常。

1. sleep(long millis)

静态方法,为当前执行线程休眠(暂时停止执行)指定毫秒数,具体取决于系统计时器和调度程序的精度和准确性,接收 long 入参,作为最大等待时间的毫秒数。在此期间此线程对象如果是中断状态,则其中断状态将被清除,并抛出 InterruptedException 异常,代码如下:

```java
//第 2 章/six/SleepTest.java
public class SleepTest {
    public static void main(String[] args) throws InterruptedException {
        Thread thread = new Thread(new Runnable() {
            @Override
            public void run() {
                try {
                    System.out.println("start");
                    Thread.sleep(3000);                //睡眠 3s
                    System.out.println("正常执行");
                } catch (InterruptedException e) {
                    System.out.println("中断");
                    System.out.println(Thread.currentThread()
                                            .isInterrupted());
                    e.printStackTrace();               //打印异常信息
                }
            }
        });
        thread.start();                                //启动线程

        Thread.sleep(1000);                            //睡眠 1s
        thread.interrupt();                            //中断此线程对象

    }
}
```

执行结果如下:

```
start
中断
```

```
false
java.lang.InterruptedException: sleep interrupted
    at java.base/java.lang.Thread.sleep(Native Method)
    at cn.kungreat.book.two.six.SleepTest$1.run(SleepTest.java:10)
    at java.base/java.lang.Thread.run(Thread.java:833)
```

2. join()

等待此线程对象销毁,在此期间等待的线程对象如果是中断状态,则其中断状态将被清除,并抛出 InterruptedException 异常,代码如下:

```java
//第 2 章/six/JoinTest.java
public class JoinTest {

    public static void main(String[] args) throws InterruptedException {
        Thread thread = new Thread(new Runnable() {
            @Override
            public void run() {
                try {
                    System.out.println("start");
                    Thread.sleep(3000);                    //睡眠 3s
                    System.out.println("正常执行");
                } catch (InterruptedException e) {
                    System.out.println("中断");
                    System.out.println(Thread.currentThread()
                                            .isInterrupted());
                    e.printStackTrace();
                }
            }
        });
        thread.start();                                    //启动线程

        Thread.currentThread().interrupt();                //主线程中断
        thread.join();                                     //主线程等待另一个线程对象销毁
        System.out.println("main-end");
    }
}
```

执行结果如下:

```
Exception in thread "main" java.lang.InterruptedException
    at java.base/java.lang.Object.wait(Native Method)
    at java.base/java.lang.Thread.join(Thread.java:1304)
    at java.base/java.lang.Thread.join(Thread.java:1372)
    at cn.kungreat.book.two.six.JoinTest.main(JoinTest.java:23)
start
正常执行
```

3. wait()

使当前执行线程等待,直到它被唤醒,通常是通知或中断。此方法需要配合 synchronized

锁使用，后续章节会详细介绍。

小结

本章详细介绍了 ThreadGroup 线程组、Thread 线程，通过本章的学习，读者需要理解它们之间的关联关系，并再次理解线程对象和当前执行线程对象的区别。在计算机的世界中，线程就是调度 CPU 去执行代码，线程跟代码并没有强的关联关系，但是在 Java 的世界中面向对象的设计概念，把线程也抽象成了一种类、对象概念。

习题

1. 判断题

(1) 线程组对象中都包含一个上级线程组对象。(   )

(2) 线程组对象可以包含多个上级线程组对象。(   )

(3) 线程组对象可以包含多个下级线程组对象。(   )

(4) 线程组对象必须有一个所管理的线程对象。(   )

(5) 线程对象必须有一个所归属的线程组对象。(   )

(6) Thread.interrupted()会清除当前中断标记。(   )

(7) 线程对象的 interrupt()，在 start()之前和之后都可以正常使用。(   )

2. 选择题

(1) 在线程组对象中，JDK 17 官方已弃用的方法有(   )。(多选)

  A. suspend() B. stop() C. destroy() D. activeCount()

(2) 为当前执行线程对象睡眠指定毫秒数的方法是(   )。(单选)

  A. Thread.sleep(long millis) B. Thread 对象.join(long millis)

  C. 线程对象.wait(long timeoutmillis) D. 线程对象.notify()

(3) Thread 类中有降低 CPU 调度效果的方法有(   )。(多选)

  A. 线程对象.setPriority(int newPriority) B. Thread.yield()

  C. 线程对象.getPriority() D. 线程对象.interrupt()

3. 填空题

查看执行结果，并补充代码，代码如下：

```
//第 2 章/answer/WorkTest.java
public class WorkTest {
    public static void main(String[] args) throws InterruptedException {
        ThreadGroup threadGroup = new ThreadGroup("A 组");
        Thread thread = _____
        thread.start();
        Thread.sleep(300);
```

```
            Thread[] threads = new Thread[threadGroup.activeCount()];
            threadGroup.enumerate(threads);
            System.out.println(Arrays.toString(threads));
    }

    static final class MyRunnable implements Runnable{

        @Override
        public void run() {
            try {
                Thread.sleep(2000);
            } catch (InterruptedException e) {
                throw new RuntimeException(e);
            }
        }
    }
}
```

执行结果如下：

```
[Thread[A 线程,5,A 组]]
```

第 3 章 多线程特性

本章介绍线程之间的协作,synchronized 对象锁、线程死锁的产生、Object 对象监视器、线程等待机制、线程唤配机制、线程可见性重排序、线程生命周期状态。

3.1 引出 synchronized 对象锁

前 2 章的课程详细介绍了 Thread 类、ThreadGroup 类,以及它们常用的方法和它们之间的关联关系。在并发环境下线程之间的协作是必不可少的,在此前的课程中基本上没有运用线程之间的协作,线程对象.join()方法是线程协作的其中一种方式,但是它的使用场景较固定,不是很灵活。

在介绍 synchronized 对象锁之前,先参考一个单线程的案例,代码如下:

8min

11min

```java
//第 3 章/one/CooperationRunnable.java
public class CooperationRunnable implements Runnable{

    public int num;
    @Override
    public void run() {
        for(int x = 0;x < 100000;x++){
            num++;
        }
    }
}
```

OneMain 类主方法,代码如下:

```java
//第 3 章/one/OneMain.java
public class OneMain {

    public static void main(String[] args) throws InterruptedException {
        CooperationRunnable cooperationRunnable = new CooperationRunnable();
        Thread thread1 = new Thread(cooperationRunnable, "A");
```

```java
        thread1.start();                //启动线程
        thread1.join();                 //等待此线程对象销毁
        System.out.println(cooperationRunnable.num);
    }
}
```

执行结果如下：

```
100000
```

单线程的执行结果，肯定是没有问题的，修改 OneMain 类，代码如下：

```java
//第 3 章/one/OneMain.java
public class OneMain {

    public static void main(String[] args) throws InterruptedException {
        CooperationRunnable cooperationRunnable = new CooperationRunnable();

        Thread thread1 = new Thread(cooperationRunnable, "A");
        thread1.start();                //启动线程

        Thread thread2 = new Thread(cooperationRunnable, "B");
        thread2.start();                //启动线程

        Thread thread3 = new Thread(cooperationRunnable, "C");
        thread3.start();                //启动线程

        thread1.join();                 //等待此线程对象销毁
        thread2.join();                 //等待此线程对象销毁
        thread3.join();                 //等待此线程对象销毁
        System.out.println(cooperationRunnable.num);
    }
}
```

多次执行代码后，观察结果。会发现每次的结果基本上都不相同，主要原因是 num++ 并非一个原子操作，如图 3-1 所示。

在多线程并发执行的情况下，可能出现例如 A 执行线程获得 num 值后，丢失了 CPU 的调度权，等到再次获得 CPU 调度权时，由于多线程并发执行的原因，此时的 num 值可能已经发生了变化，但是 A 执行线程使用的 num 值还是之前获得的旧值，在这种情况下就造成了数据的脏读，A 执行线程继续把 num 值增加 1，然后设置新的 num 值，这时又造成了数据的脏写或者覆盖。

可以使用 synchronized 对象锁来解决此问题，修改 CooperationRunnable 类，代码如下：

第 3 章 多线程特性

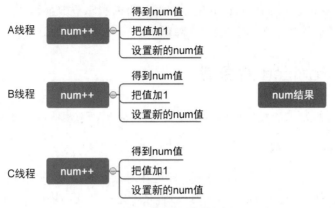

图 3-1 并发产生的脏读

```
//第 3 章/one/CooperationRunnable.java
public class CooperationRunnable implements Runnable{

    public int num;
    @Override
    public void run() {
        for(int x = 0;x < 100000;x++){
            synchronized (this){ //同一个对象锁,同时只能有一个线程进入执行
                num++;
            }
        }
    }
}
```

运行 OneMain 类主方法,执行结果如下:

```
300000
```

synchronized 对象锁,在锁同一个对象的情况下,同时只能有一个线程进入执行,其余阻塞等待拿锁,拿锁的过程是非公平的,并不是说阻塞等待时间越长拿锁概率越高,如图 3-2 所示。

图 3-2 阻塞等待拿锁

在使用 synchronized 对象锁时,需要注意锁的力度越小范围越好,毕竟是希望多线程并发执行提高效率,如果锁的范围过大,则会降低多线程并发执行的效果。

3.2 synchronized 对象锁

synchronized 对象锁按对象类型可划分为两类,即标准对象、class 对象。

3.2.1 标准对象

使用方式,可以把 synchronized 关键字标记在对象的方法上,此方法内所有代码块属于锁范围,也可以使用 synchronized(标准对象)直接锁定一个代码块区域,代码如下:

14min

7min

8min

6min

```java
//第 3 章/two/CooperationTwo.java
public class CooperationTwo implements Runnable{
    public int num;

    @Override
    public void run() {
        for(int x = 0;x < 100000;x++){
            synchronized (this){ //同一个锁对象,同时只能有一个线程进入执行
                num++;
            }
        }
    }

    public void run1(){
        for( int x = 0;x < 100000;x++){
            run2(); //同一个锁对象,同时只能有一个线程进入执行
        }
    }
    //synchronized 关键字标记在对象的方法上
    public synchronized void run2(){
      /* synchronized (this){
            num++;
        } */
        num++;
    }
}
```

TwoMain 类主方法,代码如下:

```java
//第 3 章/two/TwoMain.java
public class TwoMain {
    public static void main(String[] args) throws InterruptedException {
        CooperationTwo cooperationTwo = new CooperationTwo();
        Thread thread1 = new Thread(cooperationTwo, "A");
        thread1.start();                //启动线程
```

```java
            Thread thread2 = new Thread(new Runnable() {
                @Override
                public void run() {
                    cooperationTwo.run1();
                }
            });
            thread2.start();            //启动线程

            for(int x = 0;x < 100000;x++){
                synchronized (cooperationTwo){
                                        //同一个锁对象,同时只能有一个线程进入执行
                    cooperationTwo.num++;
                }
            }
            thread1.join();             //等待 thread1 线程对象销毁
            thread2.join();             //等待 thread2 线程对象销毁

            System.out.println(cooperationTwo.num);
    }
}
```

执行结果如下:

```
300000
```

注意:还是同一个核心概念,只是使用方式不同。synchronized 对象锁,在锁同一个对象的情况下,同时只能有一个线程进入执行,其余阻塞等待拿锁,拿锁的过程是非公平的,并不是说阻塞等待时间越长,拿锁概率越高。

3.2.2 class 对象

使用方式,可以把 synchronized 关键字标记在静态方法上,此方法内的所有代码块属于锁范围,也可以使用 synchronized(class 对象)直接锁定一个代码块区域,代码如下:

```java
//第 3 章/two/CooperationTwo.java
public class CooperationTwo implements Runnable{
    public static int num;

    @Override
    public void run() {
        for(int x = 0;x < 100000;x++){
            synchronized (CooperationTwo.class){
                        //同一个锁对象,同时只能有一个线程进入执行
                num++;
            }
        }
    }
```

```java
    }
    public void run1(){
        for(int x = 0;x < 100000;x++){
            run2(); //同一个锁对象,同时只能有一个线程进入执行
        }
    }

    public synchronized static void run2(){
        num++;
    }
}
```

TwoMain 类主方法,代码如下:

```java
//第 3 章/two/TwoMain.java
public class TwoMain {
    public static void main(String[] args) throws InterruptedException {
        CooperationTwo cooperationTwo = new CooperationTwo();
        Thread thread1 = new Thread(cooperationTwo, "A");
        thread1.start();              //启动线程

        Thread thread2 = new Thread(new Runnable() {
            @Override
            public void run() {
                cooperationTwo.run1();
            }
        },"B");
        thread2.start();              //启动线程

        for(int x = 0;x < 100000;x++){
            synchronized (cooperationTwo){
                                      //同一个锁对象,同时只能有一个线程进入执行
                CooperationTwo.num++;
            }
        }

        thread1.join();               //等待 thread1 线程对象销毁
        thread2.join();               //等待 thread2 线程对象销毁

        System.out.println(CooperationTwo.num);
    }
}
```

运行 TwoMain 类主方法,并观察执行结果,会发现结果基本上每次都不同,这是因为上面的代码有 3 个线程并行,使用了两把锁,A、B 线程使用了 CooperationTwo.class 对象锁,但是主线程使用的是 cooperationTwo 标准对象锁。修改 TwoMain 类,代码如下:

```java
//第 3 章/two/TwoMain.java
public class TwoMain {
```

```java
    public static void main(String[] args) throws InterruptedException {
        CooperationTwo cooperationTwo = new CooperationTwo();
        Thread thread1 = new Thread(cooperationTwo, "A");
        thread1.start();                    //启动线程

        Thread thread2 = new Thread(new Runnable() {
            @Override
            public void run() {
                cooperationTwo.run1();
            }
        },"B");
        thread2.start();                    //启动线程

        for(int x = 0;x < 100000;x++){
            synchronized (CooperationTwo.class){
                                            //同一个锁对象,同时只能有一个线程进入执行
                CooperationTwo.num++;
            }
        }

        thread1.join();                     //等待 thread1 线程对象销毁
        thread2.join();                     //等待 thread2 线程对象销毁

        System.out.println(CooperationTwo.num);
    }
}
```

执行结果如下:

```
300000
```

3.2.3 锁特性

synchronized 对象锁具有的特性:异常自动释放锁、可重入、不响应中断、非公平锁。

1. 异常自动释放锁

拿锁、释放锁是一个关联性很强的逻辑,在正常情况下,拿锁后进入代码块执行,出了代码块就会自动释放锁。如果出现异常情况,则锁不能自动释放是一个很严重的问题,可以想象等待拿锁的线程永远拿不到锁,相当于这一块业务无法处理而造成线程阻塞,甚至引起整个应用软件的瘫痪。好在 synchronized 对象锁有异常自动释放的特性,代码如下:

```java
//第 3 章/two/TwoMain2.java
public class TwoMain2 {
    public static void main(String[] args) throws InterruptedException {
        Runnable runnable = new Runnable() {
            @Override
            public void run() {
                synchronized (this){
```

```java
            System.out.println(Thread.currentThread().getName());
            try {
                Thread.sleep(6000);
            } catch (InterruptedException e) {
                throw new RuntimeException(e);
            }
            int num = 10 / 0;                   //模拟异常情况
            System.out.println(Thread.currentThread().getName()
                    + " - end");
        }
    }
};
Thread thread = new Thread(runnable,"A");
thread.start();                             //启动线程

Thread.sleep(100);
synchronized (runnable){
    System.out.println(Thread.currentThread().getName());
}
```

执行结果如下：

```
A
main
Exception in thread "A" java.lang.ArithmeticException: / by zero
    at cn.kungreat.book.three.two.TwoMain2$1.run(TwoMain2.java:15)
    at java.base/java.lang.Thread.run(Thread.java:833)
```

2. 可重入

相同的对象锁，在持有锁时可以直接进入，代码如下：

```java
//第3章/two/TwoMain3.java
public class TwoMain3 {

    public static void main(String[] args) throws InterruptedException {
        Runnable runnable = new Runnable() {

            @Override
            public void run() {
                synchronized (this){
                    try {
                        Thread.sleep(3000);
                    } catch (InterruptedException e) {
                        throw new RuntimeException(e);
                    }
                    System.out.println(Thread.currentThread().getName());
                    reentry();              //可重入,同一个锁对象
                }
            }
```

```
        public synchronized void reentry(){
            System.out.println(Thread.currentThread().getName()
                            + ":reentry");
        }
    };
    Thread thread = new Thread(runnable,"A");
    thread.start();

    Thread.sleep(100);              //睡眠 100ms
    synchronized (runnable){
        System.out.println(Thread.currentThread().getName());
    }
}
```

执行结果如下：

```
A
A:reentry
main
```

3. 不响应中断

synchronized 对象锁不响应中断。如果要用到中断，则需要自己设计实现。

4. 非公平锁

由系统调度分配资源，并不是等待拿锁时间越久拿锁概率越高，所以说它是非公平的。

3.3 线程死锁的产生

由于 synchronized 对象锁的特性，以下代码会产生死锁，代码如下：

 13min

```
//第 3 章/three/ThreeMain.java
public class ThreeMain {

    public static void main(String[] args) {
        DeadRunnable deadRunnable = new DeadRunnable();
        Thread thread = new Thread(deadRunnable,"A");
        thread.start();

        deadRunnable.deadLock();
    }

    static class DeadRunnable implements Runnable{
        private Object objA = new Object();         //objA 对象锁
        private Object objB = new Object();         //objB 对象锁

        @Override
```

```java
        public void run() {
            synchronized (objA){
                System.out.println(Thread.currentThread().getName());
                try {
                    Thread.sleep(1000);
                } catch (InterruptedException e) {
                    throw new RuntimeException(e);
                }
                deadLock();
            }
        }

        public void deadLock(){
            synchronized (objB){
                System.out.println(Thread.currentThread().getName());
                try {
                    Thread.sleep(1000);
                } catch (InterruptedException e) {
                    throw new RuntimeException(e);
                }
                run();
            }
        }
    }
```

> **注意**：以上代码只为了演示死锁，其本身也存在相互调用的缺陷。A 线程拿锁 objA 后再去调 deadLock() 方法，阻塞等待拿锁 objB。main 线程拿锁 objB 后再去调 run() 方法，阻塞等待拿锁 objA。

可以通过官方提供的工具 JConsole、jstack 检测死锁。

3.3.1　JConsole

在命令行直接运行 JConsole，如图 3-3 所示。
选择自己要查看的类，如图 3-4 所示。
连接以后，选择线程，并单击"检测死锁"按钮，如图 3-5 所示。
查看死锁情况，如图 3-6 所示。

3.3.2　jstack

通过 jps 查看 pid 信息，如图 3-7 所示。
使用 jstack-l pid，选择自己要查看的 pid 信息，如图 3-8 所示。
jstack 会打印出 pid 下所有的堆栈信息，内容比较多，这里关注核心的点，如图 3-9 所示。

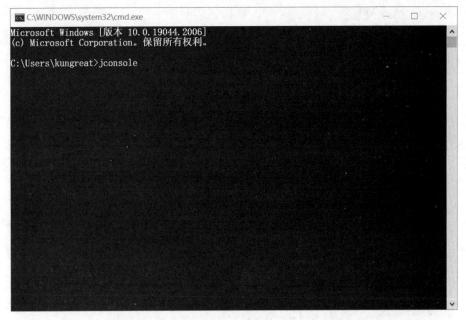

图 3-3 命令行

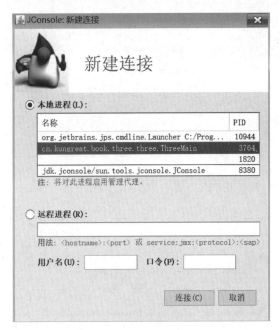

图 3-4 JConsole 首页

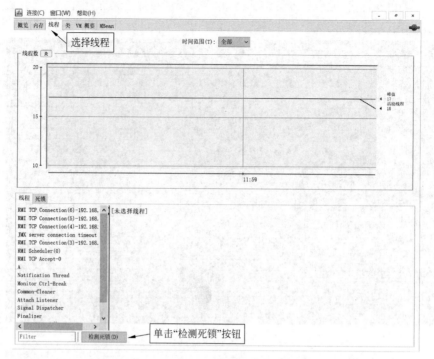

图 3-5 检测死锁

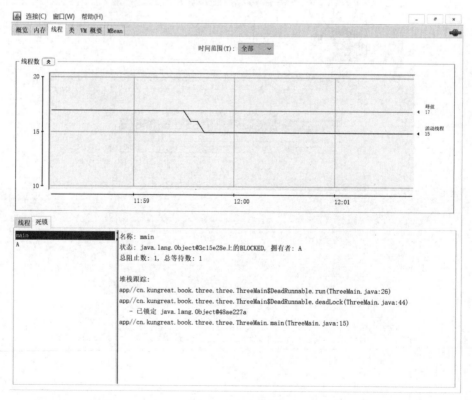

图 3-6 查看死锁

图 3-7　查看 pid

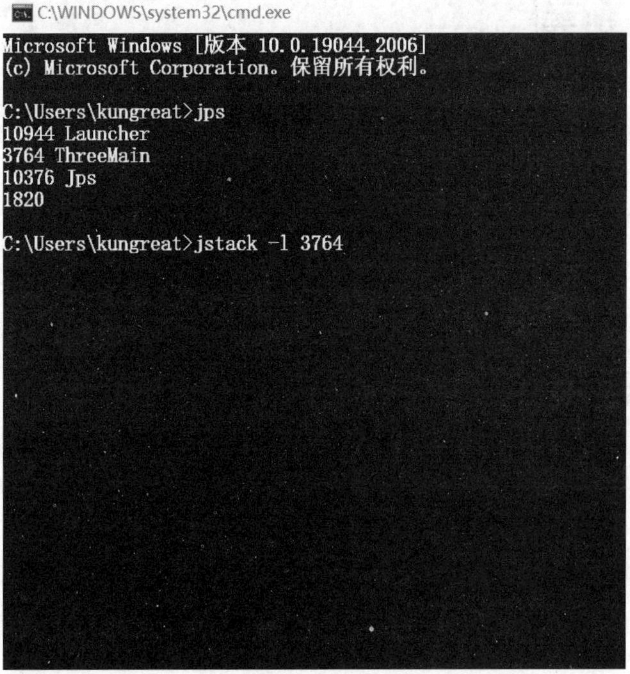

图 3-8　jstack 查看 pid

```
C:\WINDOWS\system32\cmd.exe

Found one Java-level deadlock:
"main":
  waiting to lock monitor 0x000001fc32056a70 (object 0x000000070f600000, a java.lang.Object),
  which is held by "A"
"A":
  waiting to lock monitor 0x000001fc32055ff0 (object 0x000000070f600098, a java.lang.Object),
  which is held by "main"
Java stack information for the threads listed above:
"main":
        at cn.kungreat.book.three.three.ThreeMain$DeadRunnable.run(ThreeMain.java:26)
        - waiting to lock <0x000000070f600000> (a java.lang.Object)
        at cn.kungreat.book.three.three.ThreeMain$DeadRunnable.deadLock(ThreeMain.java:44)
        - locked <0x000000070f600098> (a java.lang.Object)
        at cn.kungreat.book.three.three.ThreeMain.main(ThreeMain.java:15)
"A":
        at cn.kungreat.book.three.three.ThreeMain$DeadRunnable.deadLock(ThreeMain.java:38)
        - waiting to lock <0x000000070f600098> (a java.lang.Object)
        at cn.kungreat.book.three.three.ThreeMain$DeadRunnable.run(ThreeMain.java:32)
        - locked <0x000000070f600000> (a java.lang.Object)
        at java.lang.Thread.run(java.base@17.0.3.1/Thread.java:833)

Found 1 deadlock.
```

图 3-9 堆栈信息

3.4 对象监视器

对象监视器就是 synchronized 对象锁中锁定的那个对象，官方文档将其称为对象监视器。对象监视器提供了几个核心的功能，如 wait、notify、notifyAll。

3.4.1 wait()

使当前执行线程阻塞等待，直到它被唤醒或中断，具有释放当前锁的特性。如果没有被唤醒或中断，则一直阻塞等待，代码如下：

```java
//第 3 章/four/FourMain.java
public class FourMain {
    public static void main(String[] args) throws Exception {
        MonitorRunnable monitorRunnable = new MonitorRunnable();
        Thread thread = new Thread(monitorRunnable, "A");
        thread.start();                        //启动线程
        Thread.sleep(100);                     //睡眠 100ms
        synchronized (monitorRunnable){
            System.out.println(Thread.currentThread().getName());
        }
        System.out.println(Thread.currentThread().getName() + ":end");
    }

    static class MonitorRunnable implements Runnable{
```

```java
        @Override
        public void run() {
            synchronized (this){
                System.out.println(Thread.currentThread().getName());
                try {
                    Thread.sleep(5000);              //睡眠 5s
                    this.wait();     //使当前线程阻塞等待,直到它被唤醒或中断[释放当前锁]
                    System.out.println("被唤醒了");
                } catch (InterruptedException e) {
                    throw new RuntimeException(e);
                }
                System.out.println(Thread.currentThread().getName() + ":end");
            }
        }
    }
```

执行结果如下:

```
A
main
main:end
```

注意:以上代码 A 线程将会一直阻塞等待。

3.4.2　wait(long timeoutMillis)

使当前执行线程阻塞等待,直到它被唤醒、中断或者超过最大等待时间,具有释放当前锁的特性,接收 long 入参,作为最大等待时间的毫秒数,代码如下:

```java
//第 3 章/four/FourMain.java
public class FourMain {
    public static void main(String[] args) throws InterruptedException {
        MonitorRunnable monitorRunnable = new MonitorRunnable();
        Thread thread = new Thread(monitorRunnable, "A");
        thread.start();
        Thread.sleep(100);                //睡眠 100ms
        synchronized (monitorRunnable){
            System.out.println(Thread.currentThread().getName());
        }
        System.out.println(Thread.currentThread().getName() + ":end");
    }

    static class MonitorRunnable implements Runnable{

        @Override
        public void run() {
            synchronized (this){
```

```java
            System.out.println(Thread.currentThread().getName());
            try {
                Thread.sleep(5000);              //睡眠5s
                this.wait(5000);
                Thread.sleep(1000);              //睡眠1s
                System.out.println("被唤醒了");
            } catch (InterruptedException e) {
                throw new RuntimeException(e);
            }
        }
        System.out.println(Thread.currentThread().getName() + ":end");
    }
}
```

在经过一定量的实时时间后，A线程将正常结束生命周期，执行结果如下：

```
A
main
main:end
被唤醒了
A:end
```

3.4.3 notify()

唤醒正在等待此对象监视器上的单个执行线程，代码如下：

```java
//第3章/four/FourMain.java
public class FourMain {

    public static void main(String[] args) throws InterruptedException {
        MonitorRunnable monitorRunnable = new MonitorRunnable();
        Thread thread = new Thread(monitorRunnable, "A");
        thread.start();
        Thread.sleep(100);                       //睡眠100ms
        synchronized (monitorRunnable){
            System.out.println(Thread.currentThread().getName());
            monitorRunnable.notify();            //唤醒正在等待此对象监视器上的单个线程
            System.out.println("唤醒正在等待此对象监视器上的单个线程");
        }

    }

    static class MonitorRunnable implements Runnable{

        @Override
        public void run() {
            synchronized (this){
                System.out.println(Thread.currentThread().getName());
```

```
            try {
                Thread.sleep(5000);
                this.wait();
                System.out.println("被唤醒了");
            } catch (InterruptedException e) {
                throw new RuntimeException(e);
            }
        }
        System.out.println(Thread.currentThread().getName() + ":end");
    }
}
```

执行结果如下：

```
A
main
唤醒正在等待此对象监视器上的单个线程
被唤醒了
A:end
```

注意：观察上方输出结果的第 3 行和第 4 行，并思考它们之间的先后顺序是怎么产生的。

当同一个对象监视器有多个执行线程 wait() 时，只会唤醒其中的一个执行线程，代码如下：

```
//第3章/four/FourMain.java
public class FourMain {

    public static void main(String[] args) throws InterruptedException {
        MonitorRunnable monitorRunnable = new MonitorRunnable();
        Thread thread = new Thread(monitorRunnable, "A");
        thread.start();
        new Thread(monitorRunnable,"B").start();

        Thread.sleep(2000);              //睡眠2s,保证A、B线程都能进入wait()
        synchronized (monitorRunnable){
            System.out.println(Thread.currentThread().getName());
            monitorRunnable.notify();    //唤醒正在等待此对象监视器上的单个线程
            System.out.println("唤醒正在等待此对象监视器上的单个线程");
        }

    }

    static class MonitorRunnable implements Runnable{

        @Override
        public void run() {
```

```java
        synchronized (this){
            System.out.println(Thread.currentThread().getName());
            try {
                this.wait();      //使当前线程阻塞等待,直到它被唤醒或中断[释放当前锁]
                System.out.println(Thread.currentThread().getName()
                        + "被唤醒了");
            } catch (InterruptedException e) {
                throw new RuntimeException(e);
            }
            System.out.println(Thread.currentThread().getName() + ":end");
        }
    }
}
```

注意:多次运行上方代码后,观察结果。

3.4.4 notifyAll()

6min

唤醒正在等待此对象监视器上的所有执行线程,代码如下:

```java
//第 3 章/four/FourMain.java
public class FourMain {

    public static void main(String[] args) throws InterruptedException {
        MonitorRunnable monitorRunnable = new MonitorRunnable();
        Thread thread = new Thread(monitorRunnable, "A");
        thread.start();
        new Thread(monitorRunnable,"B").start();

        Thread.sleep(2000);                //睡眠 2s,保证 A、B 线程都能进入 wait()
        synchronized (monitorRunnable){
            System.out.println(Thread.currentThread().getName());
            monitorRunnable.notifyAll();
            System.out.println("唤醒正在等待此对象监视器上的所有线程");
        }

    }

    static class MonitorRunnable implements Runnable{

        @Override
        public void run() {
            synchronized (this){
                System.out.println(Thread.currentThread().getName());
                try {
                    this.wait();      //使当前线程阻塞等待,直到它被唤醒或中断[释放当前锁]
                    System.out.println(Thread.currentThread().getName()
                            + "被唤醒了");
```

```
            } catch (InterruptedException e) {
                throw new RuntimeException(e);
            }
        }
        System.out.println(Thread.currentThread().getName() + ":end");
    }
}
```

执行结果如下:

```
A
B
main
唤醒正在等待此对象监视器上的所有线程
A 被唤醒了
B 被唤醒了
B:end
A:end
```

对象监视器的等待、唤醒机制配合使用可以很大程度地提高程序的灵活性,此处是模拟生产和消费的案例,代码如下:

12min

```
//第 3 章/four/ConsumeMain.java
public class ConsumeMain {
    public static void main(String[] args) {
        ConsumeRunnable consumeRunnable = new ConsumeRunnable();
        new Thread(consumeRunnable,"A").start();
        consumeRunnable.consume();
    }

    static class ConsumeRunnable implements Runnable{
        private Boolean consume = false;           //生产、消费状态标识
        private Object data = null;                //存放消费的数据

        @Override
        public void run() {
            for (int x = 0;x < 10;x++){
                synchronized (this){
                    if(consume){
                        try {
                            this.wait();
                            Thread.sleep(3000);
                        } catch (InterruptedException e) {
                            throw new RuntimeException(e);
                        }
                    }
                    data = x;
                    consume = true;
                    this.notify();                 //有没有线程 wait 都可以调用
```

```java
                    System.out.println(Thread.currentThread().getName()
                            +":生产了" + x);
                }
            }
        }

        public void consume(){
            for(int x = 0;x < 10;x++){
                synchronized (this){
                    if(!consume){
                        try {
                            this.wait();
                        } catch (InterruptedException e) {
                            throw new RuntimeException(e);
                        }
                    }
                    System.out.println(Thread.currentThread().getName()
                            +":消费了" + data);
                    data = null;
                    consume = false;
                    this.notify();            //有没有线程 wait 都可以调用
                }
            }
        }
    }
```

执行结果如下：

```
A:生产了 0
main:消费了 0
A:生产了 1
main:消费了 1
A:生产了 2
main:消费了 2
A:生产了 3
main:消费了 3
A:生产了 4
main:消费了 4
A:生产了 5
main:消费了 5
A:生产了 6
main:消费了 6
A:生产了 7
main:消费了 7
A:生产了 8
main:消费了 8
A:生产了 9
main:消费了 9
```

注意：notify()唤醒线程后，此线程一样需要执行拿锁的过程。

3.5 线程的可见性和重排序

可见性和重排序跟 CPU、系统、Java 底层的设计相关，这里主要关注怎么使用。可见性表示的是当某个线程修改了数据时，对于其他线程，是否能够立即感知数据被修改了。重排序表示的是在代码编译期间，对代码进行了顺序的改变。

3.5.1 可见性

在极端的情况下，某个线程修改了数据，对其他线程不可见，代码如下：

8min

```java
//第 3 章/five/FiveMain.java
public class FiveMain {
    static boolean running = true;              //下次会修改的代码

    public static void main(String[] args) throws InterruptedException {
        Thread thread = new Thread(new Runnable() {
            @Override
            public void run() {
                while (running){ //极端的一种演示情况

                }
                System.out.println(Thread.currentThread().getName()
                                              +":end");
            }
        },"A");
        thread.start();
        Thread.sleep(100);
        running = false;                        //修改数据
        System.out.println(Thread.currentThread().getName() + ":end");
    }
}
```

执行结果如下：

```
main:end
```

注意：由于 A 线程并没有感知到 running 数据被修改了，所以一直没有结束。

修改 FiveMain 类，代码如下：

```java
//第 3 章/five/FiveMain.java
public class FiveMain {
    static volatile boolean running = true;
```

```
public static void main(String[] args) throws InterruptedException {
    Thread thread = new Thread(new Runnable() {
        @Override
        public void run() {
            while (running){ //极端的一种演示情况
            
            }
            System.out.println(Thread.currentThread().getName()
                                                    + ":end");
        }
    },"A");
    thread.start();
    Thread.sleep(100);
    running = false;
    Thread.sleep(100);
    System.out.println(Thread.currentThread().getName() + ":end");
}
```

执行结果如下：

```
A: end
main: end
```

使用 volatile 关键字可以解决可见性问题。

3.5.2 重排序

7min

表示的是在代码编译期间，对代码进行了顺序的改变。很难真实地验证，大家理解概念即可，代码如下：

2min

```
//第 3 章/five/PossibleReordering.java
public class PossibleReordering {
    static int x, y = 0;              //下次会修改的代码
    static int a, b = 0;              //下次会修改的代码
    
    public static void main(String[] args) throws Exception {
        for (int num = 0;;) {
            num++;
            x = 0;y = 0;a = 0;b = 0;       //重置数据
            Thread threadA = new Thread(new Runnable() {
                @Override
                public void run() {
                    a = 1;
                    x = b;
                    /* 重排序后
                     * x = b;
                     * a = 1;
                     */
```

```
        }
    },"A");
    Thread threadB = new Thread(new Runnable() {
        @Override
        public void run() {
            b = 1;
            y = a;
        }
    },"B");
    threadA.start();            //启动线程
    threadB.start();            //启动线程
    threadA.join();
    threadB.join();
    if(x == 0 && y == 0) {
        System.out.println(num);
        return;
    }
        }
    }
}
```

运行 PossibleReordering 类主方法,等待执行结果。分析上面的代码,可能产生的结果有 3 种。

(1) A 线程先执行完:x==0、y==1。
(2) B 线程先执行完:x==1、y==0。
(3) 一起执行完:x==1、y==1。

如果出现 x==0、y==0 的情况,即产生了异常数据,则可能就是由重排序造成的,如图 3-10 所示。

可以使用 volatile 关键字解决重排序问题,代码如下:

图 3-10 重排序后的效果

```
public class PossibleReordering {
    static volatile int x, y = 0;
    static volatile int a, b = 0;

    public static void main(String[] args) throws Exception {
        for (int num = 0;;) {
            num++;
            x = 0;y = 0;a = 0;b = 0;            //重置数据
            Thread threadA = new Thread(new Runnable() {
                @Override
                public void run() {
                    a = 1;
                    x = b;
                    /* 重排序后
                     * x = b;
                     * a = 1;
                     */
```

```
            }
        },"A");
        Thread threadB = new Thread(new Runnable() {
            @Override
            public void run() {
                b = 1;
                y = a;
            }
        },"B");
        threadA.start();            //启动线程
        threadB.start();            //启动线程
        threadA.join();
        threadB.join();
        if(x == 0 && y == 0) {
            System.out.println(num);
            return;
        }
    }
}
```

注意：volatile 关键字可以解决可见性、重排序的问题。思考 volatile、synchronized 之间的区别。

3.6 线程生命周期状态

Thread.State 是一个枚举类。一个线程在给定的时间点只能处于一种状态，这些状态是不反映任何操作系统线程状态的虚拟机线程状态。

3.6.1 NEW

尚未启动的线程处于此状态，代码如下：

```
//第 3 章/six/SixMain.java
public class SixMain {

    public static void main(String[] args) throws InterruptedException {
        StateRunnable stateRunnable = new StateRunnable();
        Thread thread = new Thread(stateRunnable,"A");

        System.out.println(thread.getState()); //线程状态
    }

    static class StateRunnable implements Runnable{

        @Override
```

```
        public void run() {
            System.out.println(Thread.currentThread().getName());
        }
    }
}
```

执行结果如下：

```
NEW
```

3.6.2 RUNNABLE

在执行中的线程处于此状态，代码如下：

```
//第3章/six/SixMain.java
public class SixMain {

    public static void main(String[] args) throws InterruptedException {
        StateRunnable stateRunnable = new StateRunnable();
        Thread thread = new Thread(stateRunnable,"A");
        thread.start();
        Thread.sleep(100);                        //睡眠100ms
        System.out.println(thread.getState());   //线程状态
    }

    static class StateRunnable implements Runnable{

        @Override
        public void run() {
            //模拟线程一直在执行中的状态
            while (true){

            }
        }
    }
}
```

执行结果如下：

```
RUNNABLE
```

3.6.3 BLOCKED

阻塞等待拿锁的执行线程处于此状态，代码如下：

```
//第3章/six/SixMain.java
public class SixMain {
```

```java
    public static void main(String[] args) throws InterruptedException {
        StateRunnable stateRunnable = new StateRunnable();
        Thread thread = new Thread(stateRunnable,"A");
        thread.start();
        synchronized (stateRunnable){//主线程先拿锁
            System.out.println(Thread.currentThread().getName());
            Thread.sleep(1000);                     //睡眠 1000ms
            System.out.println(thread.getState());  //线程状态
        }
    }

    static class StateRunnable implements Runnable{

        @Override
        public void run() {
            try {
                Thread.sleep(100);                  //睡眠 100ms
            } catch (InterruptedException e) {
                throw new RuntimeException(e);
            }
            synchronized (this){
                System.out.println(Thread.currentThread().getName());
            }
        }
    }
}
```

执行结果如下:

```
main
BLOCKED
A
```

3.6.4　WAITING

无限期等待的执行线程处于此状态,代码如下:

```java
//第 3 章/six/SixMain.java
public class SixMain {

    public static void main(String[] args) throws InterruptedException {
        StateRunnable stateRunnable = new StateRunnable();
        Thread thread = new Thread(stateRunnable,"A");
        thread.start();
        Thread.sleep(100);                          //睡眠 100ms
        synchronized (stateRunnable){
            System.out.println(thread.getState());  //线程状态
            stateRunnable.notify();                 //唤醒正在等待此对象监视器上的单个线程
        }
    }
```

```java
    static class StateRunnable implements Runnable{

        @Override
        public void run() {
            synchronized (this){
                try {
                    this.wait();
                    //使当前线程阻塞等待,直到它被唤醒或中断,具有释放当前锁的特性
                } catch (InterruptedException e) {
                    throw new RuntimeException(e);
                }
                System.out.println(Thread.currentThread().getName());
            }
        }
    }
}
```

执行结果如下:

```
WAITING
A
```

3.6.5　TIMED_WAITING

最大等待指定时间的执行线程处于此状态,代码如下:

```java
//第 3 章/six/SixMain.java
public class SixMain {

    public static void main(String[] args) throws InterruptedException {
        StateRunnable stateRunnable = new StateRunnable();
        Thread thread = new Thread(stateRunnable,"A");
        thread.start();
        Thread.sleep(100);                          //睡眠 100ms
        synchronized (stateRunnable){
            System.out.println(thread.getState());  //线程状态
            stateRunnable.notify();                 //唤醒正在等待此对象监视器上的单个线程
        }
    }

    static class StateRunnable implements Runnable{

        @Override
        public void run() {
            synchronized (this){
                try {
                    this.wait(2000);
                    //使当前线程阻塞等待、直到它被唤醒、中断或者直到经过一定量的实时时间
                    //具有释放当前锁的特性
```

```
            } catch (InterruptedException e) {
                throw new RuntimeException(e);
            }
            System.out.println(Thread.currentThread().getName());
        }
    }
}
```

执行结果如下：

```
TIMED_WAITING
A
```

3.6.6 TERMINATED

已退出的执行线程处于此状态，代码如下：

```
//第3章/six/SixMain.java
public class SixMain {

    public static void main(String[] args) throws InterruptedException {
        StateRunnable stateRunnable = new StateRunnable();
        Thread thread = new Thread(stateRunnable,"A");
        thread.start();
        Thread.sleep(100);                          //睡眠100ms
        System.out.println(thread.getState());      //线程状态
    }

    static class StateRunnable implements Runnable{

        @Override
        public void run() {
            System.out.println(Thread.currentThread().getName());
        }
    }
}
```

执行结果如下：

```
A
TERMINATED
```

小结

通过本章的学习，相信读者已经掌握了多线程协作的核心知识点。合理地利用这些知识点，已经足够在大多数情况下去处理多线程的并发问题。后续章节还会介绍其他协作方式,充分理解线程之间的协作概念,也是为学习后续课程奠定坚实的基础。

习题

1. 判断题

(1) Thread.sleep(2000)有释放当前锁的特性。(　　)
(2) 对象监视器.wait()有释放当前锁的特性。(　　)
(3) 对象监视器.notify()唤醒正在等待此对象监视器上的单个执行线程。(　　)
(4) volatile 关键字可以解决线程之间的可见性、重排序问题。(　　)
(5) 线程的生命周期共有 6 种状态,在给定时间点只能处于其中的一种状态。(　　)
(6) synchronized 对象锁可以锁任何对象。(　　)
(7) 对象监视器.wait(6000)可以在等待时间结束前唤醒。(　　)

2. 选择题

(1) 可以检测死锁的官方工具有(　　)。(多选)
　　A. JConsole　　　B. keytool　　　　C. jstack　　　　D. jar
(2) 获得 Thread 当前执行线程对象正确的方法是(　　)。(单选)
　　A. this
　　B. Thread.currentThread()
　　C. this.getName()
　　D. Thread.currentThread().getName()
(3) Thread 类中有降低 CPU 调度执行效果的方法有(　　)。(多选)
　　A. 线程对象.setPriority(int newPriority)　B. Thread.yield()
　　C. 线程对象.getPriority()　　　　　　　　D. 线程对象.interrupt()
(4) volatile 关键字解决的问题有(　　)。(多选)
　　A. 可见性　　　B. 并发脏读　　　C. 并发脏写　　　D. 重排序

3. 填空题

(1) 查看执行结果,并补充代码,代码如下:

```
//第 3 章/answer/SixMain.java
public class SixMain {

    private _____ static boolean running = true;
    public static void main(String[] args) throws InterruptedException {
        StateRunnable stateRunnable = new StateRunnable();
        Thread thread = new Thread(stateRunnable,"A");
        thread.start();
        Thread.sleep(100);                          //睡眠 100ms

        System.out.println(thread.getState());      //线程状态
```

```
            _____;
        }
    static class StateRunnable implements Runnable{
        @Override
        public void run() {
            while (running){

            }
            System.out.println(Thread.currentThread().getName() + ":end");
        }
    }
}
```

执行结果如下：

```
RUNNABLE
A:end
```

（2）根据业务要求补全代码，A、B两个执行线程轮流按照1～100的顺序打印数字，代码如下：

```
//第3章/answer/LoopPrint.java
public class LoopPrint {
    public static void main(String[] args) throws InterruptedException {
        LoopRunnable loopRunnable = new LoopRunnable();
        Thread threadA = new Thread(loopRunnable, ____);
        Thread threadB = new Thread(loopRunnable, ____);
        threadA.start();
        threadB.start();
    }

    static class LoopRunnable implements Runnable{
        private volatile int num = 1;
        @Override
        public void run() {
            for (;num <= 100;){
                synchronized (this){
                    System.out.println(Thread.currentThread().getName()
                    + ":" + num);
                    num++;
                    _____;
                    try {
                        _____;
                    } catch (InterruptedException e) {
                        throw new RuntimeException(e);
                    }
                }
            }
        }
    }
}
```

```
            }
            synchronized (this){
                _____;
            }
        }
    }
}
```

(3) 根据业务要求补全代码，启动 3 个执行线程 A、B、C，每个执行线程将自己的名称轮流打印 5 遍，打印顺序是 ABCABC…，代码如下：

```
//第 3 章/answer/ThreePrint.java
public class ThreePrint {

    public static void main(String[] args) {
        LoopRunnable loopRunnable = new LoopRunnable();
        new Thread(loopRunnable, "A").start();
        new Thread(loopRunnable, "B").start();
        new Thread(loopRunnable, "C").start();
    }

    static class LoopRunnable implements Runnable {
        private volatile int loopIndex = 0;
        private final String[] loopNames = {"A", "B", "C"};

        @Override
        public void run() {
            for (int x = 0; x < 5; x++) {
                synchronized (this) {
                    String name = Thread.currentThread().getName();
                    //名称不匹配时一直循环
                    while (!name.equals(loopNames[loopIndex])) {
                        try {
                            _____;
                        } catch (InterruptedException e) {
                            throw new RuntimeException(e);
                        }
                    }
                    System.out.print(name);                //消费名称
                    loopIndex++;
                    if (loopIndex == loopNames.length) {
                        loopIndex = 0;                     //重置指针
                    }
                    _____;
                }
            }
        }
    }
}
```

(4) 根据业务要求补全代码,包括生产和消费,生产的总量为 10,代码如下:

```java
//第 3 章/answer/ConsumeMain.java
public class ConsumeMain {
    public static void main(String[] args) {
        ConsumeRunnable consumeRunnable = new ConsumeRunnable();
        new Thread(consumeRunnable,"A").start();
        consumeRunnable.consume();
    }

    static class ConsumeRunnable implements Runnable{
        private Boolean consume = false;         //生产、消费状态标识
        private Object data = null;              //存放消费的数据

        @Override
        public void run() {
            for (int x = 0;x < 10;x++){
                synchronized (this){
                    if(consume){
                        try {
                            _____;
                            Thread.sleep(1000);
                        } catch (InterruptedException e) {
                            throw new RuntimeException(e);
                        }
                    }
                    data = x;
                    consume = true;
                    _____;             //有没有线程 wait 都可以调用
                    System.out.println(Thread.currentThread().getName()
                            + ":生产了" + x);
                }
            }
        }

        public void consume(){
            for(int x = 0;x < 10;x++){
                synchronized (this){
                    if(!consume){
                        try {
                            _____;
                        } catch (InterruptedException e) {
                            throw new RuntimeException(e);
                        }
                    }
                    System.out.println(Thread.currentThread().getName()
                            + ":消费了" + data);
                    data = null;
```

```
                consume = false;
                _____;  //有没有线程 wait 都可以调用
            }
        }
    }
}
```

第 4 章　ThreadLocal 线程局部变量

思考这样一个问题，有一段代码由于业务逻辑较复杂，其整个方法调用链路特别长，其中一个业务需求是在代码开始时初始化数据，然后此数据在后续方法调用中一直传递，供后续部分方法使用。如果真的在方法之间往下传递数据，则会导致代码很臃肿，而且并不是所有的方法都需要用到此数据，也会造成代码耦合度很高，从而导致后期难以维护。

ThreadLocal 对象线程局部变量在使用时，在线程之间是封闭且隔离的，并具有并发安全的特点。底层实现基于 Thread 对象，在进行数据跨链路传递时，使用 ThreadLocal 对象可以很优雅地实现。

4.1　在方法链路中传递数据

在方法链路中传递数据具有线程并发安全、代码臃肿、耦合度高、后期难维护等特点，代码如下：

```java
//第 4 章/one/OneMain.java
public class OneMain {

    public static void main(String[] args) {
        LinkRunnable linkRunnable = new LinkRunnable();
        new Thread(linkRunnable).start();
    }

    static class LinkRunnable implements Runnable{
        private static final Random RANDOM = new Random();
        /*
         * 在方法链路中传递数据具有线程并发安全、代码臃肿、耦合度高、后期难维护等问题
         */
        @Override
        public void run() {
            Map map = new HashMap<>();
            map.put("LINK",RANDOM.nextInt());
            LinkOne.one(map);
        }
```

```java
        }
}

//第 4 章/one/LinkOne.java
public class LinkOne {

    static void one(Map map){
        System.out.println(map.get("LINK"));
        LinkTwo.two(map);
    }
}

//第 4 章/one/LinkTwo.java
public class LinkTwo {
    static void two(Map map){
        LinkThree.three(map);
    }
}

//第 4 章/one/LinkThree.java
public class LinkThree {

    static void three(Map map){
        System.out.println(map.get("LINK"));
    }
}
```

4.2 引出线程局部变量

2min

使用线程局部变量,具有线程并发安全的特点,代码如下:

```java
//第 4 章/two/TwoMain.java
public class TwoMain {

    /*
     * LINK、LINKTWO 类似 Map 结构里的 key
     */
    public static final ThreadLocal<Object> LINK = new ThreadLocal<>();
    public static final ThreadLocal<Object> LINKTWO = new ThreadLocal<>();

    public static void main(String[] args) {
        LinkRunnable linkRunnable = new LinkRunnable();
        new Thread(linkRunnable).start();
    }

    static class LinkRunnable implements Runnable{
        private static final Random RANDOM = new Random();
        @Override
```

```java
        public void run() {
            LINK.set(RANDOM.nextInt());           //设置线程局部变量数据
            LINKTWO.set("LINKTWO");               //设置线程局部变量数据
            LinkOne.one();
        }
    }
}

//第 4 章/two/LinkOne.java
public class LinkOne {

    static void one(){
        System.out.println(TwoMain.LINK.get());      //获得线程局部变量数据
        System.out.println(TwoMain.LINKTWO.get());
        LinkTwo.two();
    }
}

//第 4 章/two/LinkTwo.java
public class LinkTwo {
    static void two(){
        LinkThree.three();
        System.out.println(TwoMain.LINKTWO.get());   //获得线程局部变量数据
    }
}

//第 4 章/two/LinkThree.java
public class LinkThree {

    static void three(){
        System.out.println(TwoMain.LINK.get());      //获得线程局部变量数据
    }
}
```

注意：思考 4.1 节和 4.2 节两种实现方式的优缺点。

4.3 线程局部变量核心概念

线程局部变量功能的实现需要三方参与，有两个操作入口。Thread 对象提供数据保存点，ThreadLocal 对象提供操作入口，InheritableThreadLocal 对象提供操作入口，ThreadLocalMap 对象提供对数据操作的功能。

4.3.1 Thread 对象数据保存点

线程局部变量的数据保存在 Thread 对象的 threadLocals、inheritableThreadLocals 字段中，所以数据是隔离的，具有线程并发安全的特点，如图 4-1 所示。

```
/* 与此线程相关的ThreadLocal值,此地图已维护
 * 通过ThreadLocal类 */
ThreadLocal.ThreadLocalMap threadLocals = null;

/*
 * 与此线程相关的InheritableThreadLocal值,这张地图是
 * 由InheritableThreadLocal类维护的
 */
ThreadLocal.ThreadLocalMap inheritableThreadLocals = null;
```

图 4-1 线程局部变量的数据保存点

1. threadLocals

当使用 ThreadLocal 对象作为操作入口时,数据会存放在当前执行线程对象的 threadLocals 字段里,代码如下:

```java
//第 4 章/three/ThreeMain.java
public class ThreeMain {

    public static final ThreadLocal< String > THREAD_LOCAL =
                                          new ThreadLocal<>();    //操作入口

    public static void main(String[] args) {
        THREAD_LOCAL.set("test");                          //设置线程局部变量数据
        ThreadLocalsRun threadLocalsRun = new ThreadLocalsRun();
        new Thread(threadLocalsRun, "A").start();
        Thread currentThread = Thread.currentThread();
        System.out.println(currentThread.getName()
                    + ":" + THREAD_LOCAL.get());           //获得线程局部变量数据
    }

    static class ThreadLocalsRun implements Runnable{
        @Override
        public void run() {
            Thread currentThread = Thread.currentThread();
            System.out.println(currentThread.getName()
                    + ":" + THREAD_LOCAL.get());      //获得线程局部变量数据
        }
    }
}
```

执行结果如下:

```
main:test
A:null
```

线程局部变量存放的数据结构类似于 Map < K , V >,K 为 ThreadLocal<?>类型,V 默

认为 Object 类型，可以通过泛型指定 V 的类型。

Debug 模式下查看主线程对象的 threadLocals 字段，并找到自己存放的 K、V，忽略 JVM 存放的 K、V，如图 4-2 所示。

```
> ≡ currentThread: Thread = {Thread@1} "Thread[main,5,main]"
    > (f) name: String = "main"
      (f) priority: int = 5
      (f) daemon: boolean = false
      (f) interrupted: boolean = false
      (f) stillborn: boolean = false
      (f) eetop: long = 2114408602688
      (f) target: Runnable = null
    > (f) group: ThreadGroup = {ThreadGroup@703} "java.lang.ThreadGroup[name=main,maxpri=10]"
    > (f) contextClassLoader: ClassLoader = {ClassLoaders$AppClassLoader@704}
    > (f) inheritedAccessControlContext: AccessControlContext = {AccessControlContext@816}
      (f) threadInitNumber: int = 0
    > (f) threadLocals: ThreadLocal$ThreadLocalMap = {ThreadLocal$ThreadLocalMap@817}
        (f) INITIAL_CAPACITY: int = 16
      > (f) table: ThreadLocal$ThreadLocalMap$Entry[] = {ThreadLocal$ThreadLocalMap$Entry[16]@820}
          不显示空元素
        > ≡ 3 = {ThreadLocal$ThreadLocalMap$Entry@822}
        ∨ ≡ 5 = {ThreadLocal$ThreadLocalMap$Entry@823}
          > (f) value: Object = "test"    ← V
          > (f) referent: Object = {ThreadLocal@705}    ← K
          > (f) queue: ReferenceQueue = {ReferenceQueue$Null@827}
            (f) next: Reference = null
            (f) discovered: Reference = null
          > (f) processPendingLock: Object = {Object@825}
            (f) processPendingActive: boolean = false
            (f) $assertionsDisabled: boolean = true
        > ≡ 10 = {ThreadLocal$ThreadLocalMap$Entry@824}
        (f) size: int = 3
        (f) threshold: int = 10
```

图 4-2　主线程对象 threadLocals 字段

Debug 模式下查看 A 线程对象的 threadLocals 字段，并找到自己存放的 K、V，忽略 JVM 存放的 K、V，当没有设置数据时 V 默认为空，如图 4-3 所示。

2. inheritableThreadLocals

当使用 InheritableThreadLocal 对象作为操作入口时，数据会存放在当前执行线程对象的 inheritableThreadLocals 字段里，代码如下：

```java
//第 4 章/three/ThreeShareMain.java
public class ThreeShareMain {

    public static final InheritableThreadLocal<String> THREAD_LOCAL =
                                new InheritableThreadLocal<>();
```

```
▼ ≡ currentThread: Thread  = {Thread@703} "Thread[A,5,main]"
    ▶ ⓕ name: String  = "A"
       ⓕ priority: int  = 5
       ⓕ daemon: boolean  = false
       ⓕ interrupted: boolean  = false
       ⓕ stillborn: boolean  = false
       ⓕ eetop: long  = 2038916868064
    ▶ ⓕ target: Runnable  = {ThreeMain$ThreadLocalsRun@745}
    ▶ ⓕ group: ThreadGroup  = {ThreadGroup@744} "java.lang.ThreadGroup[name=main,maxpri=10]"
    ▶ ⓕ contextClassLoader: ClassLoader  = {ClassLoaders$AppClassLoader@746}
    ▶ ⓕ inheritedAccessControlContext: AccessControlContext  = {AccessControlContext@850}
       ⓕ threadInitNumber: int  = 0
    ▼ ⓕ threadLocals: ThreadLocal$ThreadLocalMap  = {ThreadLocal$ThreadLocalMap@851}
          ⓕ INITIAL_CAPACITY: int  = 16
       ▼ ⓕ table: ThreadLocal$ThreadLocalMap$Entry[]  = {ThreadLocal$ThreadLocalMap$Entry[16]@854}
             不显示空元素
          ▼ ≡ 5 = {ThreadLocal$ThreadLocalMap$Entry@856}
                ⓕ value: Object  = null         ← [ V ]
             ▶ ⓕ referent: Object  = {ThreadLocal@864}  ← [ K ]
             ▶ ⓕ queue: ReferenceQueue  = {ReferenceQueue$Null@862}
                ⓕ next: Reference  = null
                ⓕ discovered: Reference  = null
             ▶ ⓕ processPendingLock: Object  = {Object@859}
                ⓕ processPendingActive: boolean  = false
                ⓕ $assertionsDisabled: boolean  = true
          ▶ ≡ 10 = {ThreadLocal$ThreadLocalMap$Entry@857}
          ▶ ≡ 12 = {ThreadLocal$ThreadLocalMap$Entry@858}
       ⓕ size: int  = 3
       ⓕ threshold: int  = 10
```

图 4-3 A 线程对象 threadLocals 字段

```java
public static void main(String[] args) {
    THREAD_LOCAL.set("test");                    //设置线程局部变量数据
    ThreadLocalsRun threadLocalsRun = new ThreadLocalsRun();
    new Thread(threadLocalsRun,"A").start();     //下次会修改的代码
    Thread currentThread = Thread.currentThread();
    System.out.println(currentThread.getName()
            + ":" + THREAD_LOCAL.get());          //获得线程局部变量数据
}

static class ThreadLocalsRun implements Runnable{

    @Override
    public void run() {
        Thread currentThread = Thread.currentThread();
        System.out.println(currentThread.getName()
                + ":" + THREAD_LOCAL.get());      //获得线程局部变量数据
    }
}
```

执行结果如下:

```
A:test
main:test
```

首先观察 InheritableThreadLocal 类源代码,发现它继承了 ThreadLocal 类,并且重写了几个重要的方法,从这些方法中可以清晰地看见数据存放使用了 Thread 对象的 inheritableThreadLocals 字段,如图 4-4 所示。

```java
public class InheritableThreadLocal<T> extends ThreadLocal<T> {

    public InheritableThreadLocal() {}

    protected T childValue(T parentValue) {
        return parentValue;
    }

    ThreadLocalMap getMap(Thread t) {
        return t.inheritableThreadLocals;
    }

    void createMap(Thread t, T firstValue) {
        t.inheritableThreadLocals = new ThreadLocalMap(this, firstValue);
    }
}
```

图 4-4　InheritableThreadLocal 类源代码

再次观察上方代码的执行结果,会发现执行线程 A 也获得了数据,这是由于存放在线程对象 inheritableThreadLocals 字段中的数据可以共享,如图 4-5 所示。

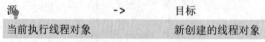

图 4-5　共享线程局部变量

共享线程局部变量的核心实现源代码在 Thread 类对象构造器上,如图 4-6 所示。

由此可以猜想,当使用 new Thread(threadLocalsRun,"A")构造器时,最终通过构造器重载默认的 inheritThreadLocals 字段传入了 true,如图 4-7 所示。

修改 ThreeShareMain 类,代码如下:

```java
//第 4 章/three/ThreeShareMain.java
public class ThreeShareMain {

    public static final InheritableThreadLocal < String > THREAD_LOCAL =
            new InheritableThreadLocal<>();
```

```java
private Thread(ThreadGroup g, Runnable target, String name,
               long stackSize, AccessControlContext acc,
               boolean inheritThreadLocals) {      ← 构造器传入
    if (name == null) {
        throw new NullPointerException("name cannot be null");
    }
    this.name = name;
    Thread parent = currentThread();    ← 当前执行线程对象
    g.addUnstarted();
    this.group = g;
    this.daemon = parent.isDaemon();
    this.priority = parent.getPriority();
    if (security == null || isCCLOverridden(parent.getClass()))
        this.contextClassLoader = parent.getContextClassLoader();
    else
        this.contextClassLoader = parent.contextClassLoader;
    this.inheritedAccessControlContext =
            acc != null ? acc : AccessController.getContext();
    this.target = target;
    setPriority(priority);         ← 构造器传入         ← 数据不为空
    if (inheritThreadLocals && parent.inheritableThreadLocals != null)
        this.inheritableThreadLocals =
            ThreadLocal.createInheritedMap(parent.inheritableThreadLocals);
    this.stackSize = stackSize;                      ← 复制数据
    this.tid = nextThreadID();
}
```

图 4-6　Thread 类对象构造器

```java
public Thread(Runnable target, String name) {
    this(null, target, name, 0);
}

public Thread(ThreadGroup group, Runnable target, String name,
              long stackSize) {
    this(group, target, name, stackSize, null, true);
}
```

图 4-7　Thread 类构造器重载

```java
public static void main(String[] args) {
    THREAD_LOCAL.set("test");                       //设置线程局部变量数据
    ThreadLocalsRun threadLocalsRun = new ThreadLocalsRun();
    new Thread(Thread.currentThread().getThreadGroup(),
            threadLocalsRun, "A", 0, false).start();
    Thread currentThread = Thread.currentThread();
    System.out.println(currentThread.getName()
            + ":" + THREAD_LOCAL.get()); //获得线程局部变量数据
}
```

```java
static class ThreadLocalsRun implements Runnable{
    @Override
    public void run() {
        Thread currentThread = Thread.currentThread();
        System.out.println(currentThread.getName()
                + ":" + THREAD_LOCAL.get());              //获得线程局部变量数据
    }
}
```

执行结果如下：

```
A:null
main:test
```

注意：在共享主线程局部变量的数据之后，再重新设置主线程的局部变量数据，会影响其他的任何线程吗？

4.3.2 线程局部变量操作入口

线程局部变量操作入口有两个，即 ThreadLocal 对象提供操作入口、InheritableThreadLocal 对象提供操作入口。

1. ThreadLocal

此对象有 4 个常用方法，提供了数据的获得、删除、设置、初始化操作。

```
protected T initialValue() {
    return null;
}
```

图 4-8 initialValue()方法

1) initialValue()

初始化数据，默认返回 null，可以继承后重写此方法，如图 4-8 所示。

2) get()

返回线程局部变量数据，调用时如果当前执行线程对象的 threadLocals 字段为空，则回调 initialValue()方法初始化数据，并且创建 ThreadLocalMap 对象，然后返回初始化数据，如图 4-9 所示。

继承 ThreadLocal 类并重写 initialValue()方法，验证初始化数据，代码如下：

```java
//第 4 章/three/ThreadLocalTest.java
public class ThreadLocalTest {

    public static final ThreadLocal<String> THREAD_LOCAL = ThreadLocal.withInitial(()->{
        System.out.println("重写 initialValue");
        return "AAA";
    }); //继承 ThreadLocal 类并重写 initialValue()方法
```

```java
public T get() {
    Thread t = Thread.currentThread();   // 当前执行线程对象
    ThreadLocalMap map = getMap(t);
    if (map != null) {
        ThreadLocalMap.Entry e = map.getEntry(this);
        if (e != null) {
            @SuppressWarnings("unchecked")
            T result = (T)e.value;
            return result;
        }
    }
    return setInitialValue();
}
private T setInitialValue() {
    T value = initialValue();   // 初始化数据
    Thread t = Thread.currentThread();
    ThreadLocalMap map = getMap(t);
    if (map != null) {
        map.set(this, value);
    } else {
        createMap(t, value);   // 创建ThreadLocalMap对象
    }
    if (this instanceof TerminatingThreadLocal) {
        TerminatingThreadLocal.register((TerminatingThreadLocal<?>)
    }
    return value;
}
```

图 4-9　threadLocals 字段为空时初始化数据

```java
public static void main(String[] args) {
    System.out.println(THREAD_LOCAL.get());   //获得线程局部变量数据

    }
}
```

执行结果如下：

重写 initialValue
AAA

3) set(T value)

设置线程局部变量的数据，接收泛型入参，作为数据的值，调用时如果当前执行线程对象的 threadLocals 字段为空，则创建 ThreadLocalMap 对象，否则设置数据（K，V），同一个对象并且同一个执行线程对象，多次设置数据将呈现覆盖效果，如图 4-10 所示。

设置线程局部变量的数据，代码如下：

```java
public void set(T value) {
    Thread t = Thread.currentThread();    // 当前执行线程对象
    ThreadLocalMap map = getMap(t);
    if (map != null) {
        map.set(this, value);
    } else {
        createMap(t, value);              // 为空时创建ThreadLocalMap对象
    }
}

ThreadLocalMap getMap(Thread t) {
    return t.threadLocals;                // 获得ThreadLocalMap对象
}

void createMap(Thread t, T firstValue) {
    t.threadLocals = new ThreadLocalMap(this, firstValue);
}
```

图 4-10　设置数据流程

```java
//第 4 章/three/ThreadLocalTest.java
public class ThreadLocalTest {
    public static final ThreadLocal<String> THREAD_LOCAL = ThreadLocal.withInitial(()->{
        System.out.println("重写 initialValue");
        return "AAA";
    });                                              //继承 ThreadLocal 类并重写 initialValue()方法

    public static void main(String[] args) {
        THREAD_LOCAL.set("main");                    //设置线程局部变量数据
        System.out.println(THREAD_LOCAL.get());      //获得线程局部变量数据
    }
}
```

执行结果如下：

```
Main
```

4) remove()

删除当前执行线程对象 threadLocals 字段中与此关联的 K、V 数据，删除点用 key 作比较，源码如图 4-11 所示。

```java
public void remove() {
    ThreadLocalMap m = getMap(Thread.currentThread());
    if (m != null) {
        m.remove(key: this);           // 删除此key的数据
    }
}
```

图 4-11　删除指定 key 数据

删除当前执行线程对象 threadLocals 字段中与此关联的 K、V 数据,代码如下:

```java
//第 4 章/three/ThreadLocalTest.java
public class ThreadLocalTest {

    public static final ThreadLocal<String> THREAD_LOCAL = ThreadLocal.withInitial(()->{
        System.out.println("重写 initialValue");
        return "AAA";
    });

    public static void main(String[] args) {
        THREAD_LOCAL.set("main");                      //设置线程局部变量数据
        Thread thread = Thread.currentThread();
        THREAD_LOCAL.remove();                          //删除此 key 数据
        System.out.println(THREAD_LOCAL.get());         //获得线程局部变量数据
    }

}
```

执行结果如下:

```
重写 initialValue
AAA
```

注意:观察上方的输出结果,会发现设置的 main 数据已被删除,最后获得的线程局部变量的数据是 initialValue() 初始化的数据。

2. InheritableThreadLocal

此类继承了 ThreadLocal 类,大部分功能的实现依赖于 ThreadLocal 类,只是重写了 3 个重要的方法,如图 4-12 所示。

```java
public class InheritableThreadLocal<T> extends ThreadLocal<T> {
    public InheritableThreadLocal() {}

    protected T childValue(T parentValue) {       ← 共享数据时需要重写的方法
        return parentValue;
    }

    ThreadLocalMap getMap(Thread t) {
        return t.inheritableThreadLocals;          ← 数据关联的线程对象字段
    }

    void createMap(Thread t, T firstValue) {
        t.inheritableThreadLocals = new ThreadLocalMap(this, firstValue);
    }                                              ← 数据关联的线程对象字段
}
```

图 4-12 InheritableThreadLocal 类源代码

4.3.3 线程局部变量数据操作功能

ThreadLocalMap 对象提供了对数据操作的功能。ThreadLocalMap 是一个静态内部类，其核心数据存放在此对象的 table 字段中，类型为 Entry[]，Entry 也是一个静态内部类，Entry 的核心功能是继承了 WeakReference，拥有了弱引用的功能，如图 4-13 所示。

```
static class ThreadLocalMap {

    static class Entry extends WeakReference<ThreadLocal<?>> {
        Object value;

        Entry(ThreadLocal<?> k, Object v) {
            super(k);  ← 弱引用垃圾回收器特别处理
            value = v;
        }
    }

    private static final int INITIAL_CAPACITY = 16;

    /**
     * 根据需要调整表格的大小，长度必须始终是2的幂
     */
    private Entry[] table;  ← 对象核心数据存放点
```

图 4-13　数据存放在 table 字段中

1. 构造器

ThreadLocalMap 构造器见表 4-1。

表 4-1　ThreadLocalMap 对象构造器

构 造 器	描 述
ThreadLocalMap(ThreadLocal<?> firstKey, Object firstValue)	构造新的对象，指定 K、V
private ThreadLocalMap(ThreadLocalMap parentMap)	构造新的对象，指定共享线程局部变量

1) ThreadLocalMap(ThreadLocal<?> firstKey，Object firstValue)

接收 ThreadLocal 入参，作为数据存放的 K，接收 Object 入参，作为数据存放的 V。类似 Map 结构，如图 4-14 所示。

此构造器在两处被使用，如图 4-15 所示。

2) private ThreadLocalMap(ThreadLocalMap parentMap)

接收 ThreadLocalMap 入参，作为共享线程局部变量数据的源，如图 4-16 所示。

此构造器只在一处被使用，如图 4-17 所示。

2. 哈希魔法值

ThreadLocal 对象中有一个 threadLocalHashCode 字段，此字段是一个哈希魔法值，在

```java
ThreadLocalMap(ThreadLocal<?> firstKey, Object firstValue) {
    table = new Entry[INITIAL_CAPACITY];  ← 初始化数组长度
    int i = firstKey.threadLocalHashCode & (INITIAL_CAPACITY - 1);
    table[i] = new Entry(firstKey, firstValue);  ← 定位数组索引存放数据
    size = 1;
    setThreshold(INITIAL_CAPACITY);  ← 设置负载因子，影响数组扩容
}
```

图 4-14　ThreadLocalMap 构造器

```java
//InheritableThreadLocal类
void createMap(Thread t, T firstValue) {
    t.inheritableThreadLocals = new ThreadLocalMap(this, firstValue);
}

//ThreadLocal类
void createMap(Thread t, T firstValue) {
    t.threadLocals = new ThreadLocalMap(this, firstValue);
}
```

图 4-15　构造器的使用

```java
private ThreadLocalMap(ThreadLocalMap parentMap) {
    Entry[] parentTable = parentMap.table;
    int len = parentTable.length;
    setThreshold(len);
    table = new Entry[len];  ← 初始化数组长度

    for (Entry e : parentTable) {
        if (e != null) {
            @SuppressWarnings("unchecked")
            ThreadLocal<Object> key = (ThreadLocal<Object>) e.get();
            if (key != null) {
                Object value = key.childValue(e.value);
                Entry c = new Entry(key, value);  ← 创建数据
                int h = key.threadLocalHashCode & (len - 1);
                while (table[h] != null)
                    h = nextIndex(h, len);  ← 定位到下一个索引
                table[h] = c;  ← 存放数据
                size++;
            }
        }
    }
}
```

图 4-16　共享线程局部变量的构造器

定位数组索引时会用到，如图 4-18 所示。

3．set(ThreadLocal<?> key，Object value)

对象私有方法，接收 ThreadLocal 入参，作为存储的 key，接收 Object 入参，作为存储的 value。在存储数据的过程中，可能发生以下 3 种情况，如图 4-19 所示。

```
//Thread类
private Thread(ThreadGroup g, Runnable target, String name,
               long stackSize, AccessControlContext acc,
               boolean inheritThreadLocals) {
    if (name == null) {
        throw new NullPointerException("name cannot be null");
    }
    this.name = name;
    Thread parent = currentThread();
    if (inheritThreadLocals && parent.inheritableThreadLocals != null)
        this.inheritableThreadLocals =
            ThreadLocal.createInheritedMap(parent.inheritableThreadLocals); ← 共享线程局部变量
}
//ThreadLocal类
static ThreadLocalMap createInheritedMap(ThreadLocalMap parentMap) {
    return new ThreadLocalMap(parentMap);
}
```

图 4-17 共享线程局部变量调用方法链路

```
public class ThreadLocal<T> {

    private final int threadLocalHashCode = nextHashCode(); ← 存储数据时，用作位运算定位数组索引
    /**
     * 并发安全
     */
    private static AtomicInteger nextHashCode =
        new AtomicInteger();

    /**
     * 连续生成的哈希代码之间的差异，将隐式顺序线程局部id转换为接近
     * 最佳分布两个大小的表的幂的乘法哈希值
     */
    private static final int HASH_INCREMENT = 0x61c88647;

    /**
     * 并发安全
     */
    private static int nextHashCode() {
        return nextHashCode.getAndAdd(HASH_INCREMENT);
    }
```

图 4-18 哈希魔法值

（1）当通过位运算定位到数组索引内的数据为空时直接存储数据。

（2）当通过位运算定位到数组索引内的数据不为空时,循环判断 key 是否相等,如果相等,则更新 value。这点类似于 Map 结构,当相同 key 多次存储数据时 value 将呈现覆盖效果。

```java
private void set(ThreadLocal<?> key, Object value) {
    Entry[] tab = table;
    int len = tab.length;
    //位运算定位数组内索引threadLocalHashCode不同也有可能定位到同一个索引
    int i = key.threadLocalHashCode & (len-1);
    //循环遍历数据
    for (Entry e = tab[i];
         e != null;
         e = tab[i = nextIndex(i, len)]) {
        if (e.refersTo(key)) {          // 情况2
            e.value = value;
            return;
        }

        if (e.refersTo(null)) {         // 情况3
            replaceStaleEntry(key, value, i);
            return;
        }
    }

    tab[i] = new Entry(key, value);     // 情况1
    int sz = ++size;
    if (!cleanSomeSlots(i, sz) && sz >= threshold)
        rehash();
}
```

图 4-19　set 方法源代码

（3）当通过位运算定位到数组索引内的数据不为空时，有可能是这个索引已经被别的数据占用，虽然每个 key 对象的 threadLocalHashCode 字段值都不相同，但是通过位运算后仍然可能定位到同一个索引，这种情况就会造成索引被占用。此时会进入循环判断，如果找到空位，则是（1）的效果，如果找到的 key 相同，则是（2）的效果，如果找到失效的数据，则替换此数据。

在存储数据的过程中，如果数据的数量超过了一定的阈值，则会有扩容的效果，如图 4-20 所示。

在存储数据的过程中，会有多次清理数据并重新排列索引的操作。清理数据的核心点是 key 等于空，而这个 key 是一个弱引用，这是一个很重要的点，如图 4-21 所示。

4. replaceStaleEntry（ThreadLocal<?> key,Object value,int staleSlot）

对象私有方法，用于替换此索引处的数据，有一个用法，在设置数据时，接收 ThreadLocal 入参，作为存储的 key，接收 Object 入参，作为存储的 value，接收 int 入参，作为数组的索引位置，如图 4-22 所示。

5. expungeStaleEntry（int staleSlot）

对象私有方法，用于清理数据并重新排列索引，接收 int 入参，作为清理数据的索引，如图 4-23 所示。

```java
private void resize() {
    Entry[] oldTab = table;
    int oldLen = oldTab.length;
    int newLen = oldLen * 2;          // 扩容后的容量
    Entry[] newTab = new Entry[newLen];   // 新建数组
    int count = 0;
    for (Entry e : oldTab) {
        if (e != null) {
            ThreadLocal<?> k = e.get();
            if (k == null) {
                e.value = null; // Help the GC
            } else {
                int h = k.threadLocalHashCode & (newLen - 1);
                while (newTab[h] != null)
                    h = nextIndex(h, newLen);
                newTab[h] = e;         // 存储数据
                count++;
            }
        }
    }
    setThreshold(newLen);              // 设置负载因子
    size = count;
    table = newTab;                    // 保存新的数组
}
```

图 4-20　数组扩容

```java
private boolean cleanSomeSlots(int i, int n) {
    boolean removed = false;
    Entry[] tab = table;
    int len = tab.length;
    do {
        i = nextIndex(i, len);
        Entry e = tab[i];
        if (e != null && e.refersTo( obj: null)) {   // key等于空
            n = len;
            removed = true;
            i = expungeStaleEntry(i);
        }                                            // 清理数据并重新排列索引
    } while ( (n >>>= 1) != 0);
    return removed;                                  // 一定次数的条件循环
}
```

图 4-21　清理数据

6. expungeStaleEntries()

对象私有方法，循环所有数据，判断 key 是否等于空，如果 key 等于空，则调用 expungeStaleEntry(int staleSlot)清理数据并重新排列索引，如图 4-24 所示。

```java
//官方源代码精简版
private void replaceStaleEntry(ThreadLocal<?> key, Object value,
                               int staleSlot) {
    Entry[] tab = table;
    int len = tab.length;
    Entry e;
    for (int i = nextIndex(staleSlot, len);
         (e = tab[i]) != null;
         i = nextIndex(i, len)) {
        if (e.refersTo(key)) {        // key相等
            e.value = value;          // 更新value
            tab[i] = tab[staleSlot];
            tab[staleSlot] = e;       // 替换索引内数据
            if (slotToExpunge == staleSlot)
                slotToExpunge = i;
            cleanSomeSlots(expungeStaleEntry(slotToExpunge), len);
            return;                   // 结束此方法
        }

        if (e.refersTo(null) && slotToExpunge == staleSlot)
            slotToExpunge = i;
    }
    tab[staleSlot].value = null;
    tab[staleSlot] = new Entry(key, value);  // 替换索引内数据
}
```

图 4-22 源代码精简版

```java
private int expungeStaleEntry(int staleSlot) {
    Entry[] tab = table;
    int len = tab.length;
    tab[staleSlot].value = null;
    tab[staleSlot] = null;            // 清理数据
    size--;
    Entry e; int i;
    for (i = nextIndex(staleSlot, len);
         (e = tab[i]) != null;
         i = nextIndex(i, len)) {
        ThreadLocal<?> k = e.get();
        if (k == null) {
            e.value = null;
            tab[i] = null;            // 清理数据
            size--;
        } else {
            int h = k.threadLocalHashCode & (len - 1);
            if (h != i) {             // 重新排列索引数据
                tab[i] = null;
                while (tab[h] != null)
                    h = nextIndex(h, len);
                tab[h] = e;
            }
        }
    }
    return i;
}
```

图 4-23 清理数据并重新排列索引

```java
private void expungeStaleEntries() {
    Entry[] tab = table;
    int len = tab.length;
    for (int j = 0; j < len; j++) {
        Entry e = tab[j];
        if (e != null && e.refersTo( obj: null))   // key等于空
            expungeStaleEntry(j);                   // 清理数据并重新排列索引
    }
}
```

图 4-24 全量清理数据

7. getEntry(ThreadLocal<?> key)

对象私有方法，用于返回 key 对应的 Entry，接收 ThreadLocal 入参，作为数据匹配的 key，由于可能出现数组索引被占用的情况，所以在当前索引匹配不上数据时，会循环往后查找匹配的数据，如图 4-25 所示。

```java
private Entry getEntry(ThreadLocal<?> key) {
    int i = key.threadLocalHashCode & (table.length - 1);   // 定位索引
    Entry e = table[i];
    if (e != null && e.refersTo(key))       // 数据匹配成功
        return e;
    else
        return getEntryAfterMiss(key, i, e);  // 往后查找匹配数据
}

private Entry getEntryAfterMiss(ThreadLocal<?> key, int i, Entry e) {
    Entry[] tab = table;
    int len = tab.length;

    while (e != null) {
        if (e.refersTo(key))            // 数据匹配成功
            return e;
        if (e.refersTo(null))           // 清理数据
            expungeStaleEntry(i);
        else
            i = nextIndex(i, len);      // 下一个索引
        e = tab[i];
    }
    return null;
}
```

图 4-25 返回 key 对应的 Entry

8. remove(ThreadLocal<?> key)

对象私有方法，用于删除指定 key 的数据，接收 ThreadLocal 入参，作为数据匹配的 key，由于可能出现数组索引被占用的情况，所以在当前索引匹配不上数据时，会循环往后查找匹配的数据，如图 4-26 所示。

```java
private void remove(ThreadLocal<?> key) {
    Entry[] tab = table;
    int len = tab.length;
    int i = key.threadLocalHashCode & (len-1);
    for (Entry e = tab[i];
         e != null;
         e = tab[i = nextIndex(i, len)]) {
        if (e.refersTo(key)) {
            e.clear();
            expungeStaleEntry(i);
            return;
        }
    }
}
```

图 4-26　删除指定 key 的数据

4.3.4　弱引用介绍

ThreadLocalMap 对象存储数据使用的是 Entry[]，Entry 是一个静态内部类，其继承了 WeakReference 类后继承了 Reference 类，存储数据的 key 最终保存在 Reference 类的 referent 对象字段中，如图 4-27 所示。

```java
static class Entry extends WeakReference<ThreadLocal<?>> {
    /** The value associated with this ThreadLocal. */
    Object value;

    Entry(ThreadLocal<?> k, Object v) {
        super(k);  // 传递 key
        value = v;
    }
}
```

图 4-27　Entry 静态内部类

Reference 类的 referent 对象字段有垃圾回收器（GC）特别处理的效果，此字段是弱引用实现的关键点，如图 4-28 所示。

```java
//Reference类
private T referent;        /* 垃圾回收器特别处理 */
```

图 4-28　垃圾回收器特别处理

在此读者回顾一下 Java 的基本垃圾回收器规则，当一个对象不再被任何字段、变量、常量引用时，此对象会被垃圾回收器回收，代码如下：

```java
//第 4 章/four/ThreadLocalMapTest.java
public class ThreadLocalMapTest {

    public static final ThreadLocal<String> THREAD_LOCAL
                                = new ThreadLocal<>();

    public static Object obj = new Object();

    public static void main(String[] args) {
        obj = null; //标准垃圾回收器,源对象所占用的内存空间将被回收

        THREAD_LOCAL.set("V");
    }
}
```

注意：THREAD_LOCAL 静态常量所引用的对象地址,在别的地方还会被引用吗? 在首次调用 set("V")方法之前或之后又有什么区别。

观察代码并思考对象地址引用发生变化的过程,代码如下:

```java
//第 4 章/four/ThreadLocalMapTest.java
public class ThreadLocalMapTest {

    public static ThreadLocal<String> THREAD_LOCAL = null;

    public static Object obj = new Object();

    public static void main(String[] args) {
        obj = null; //源对象所占用的内存空间将会被回收
        for (int x = 1; x < 6; x++){
            THREAD_LOCAL = new ThreadLocal<>();
            THREAD_LOCAL.set(String.valueOf(x));
        }
        Thread currentThread = Thread.currentThread();

        System.out.println(THREAD_LOCAL.get());
    }
}
```

执行结果如下:

5

用 Debug 模式查看线程局部变量的数据,发现 1~5 的数据都存放在 table 字段中,如图 4-29 所示。

由此可以得出结论,循环创建的 ThreadLocal 对象,前 4 个只在 table 字段中被引用,而

```
 threadInitNumber: int = 0
 threadLocals: ThreadLocal$ThreadLocalMap = {ThreadLocal$ThreadLocalMap@815}  ← Thread对象
     INITIAL_CAPACITY: int = 16
     table: ThreadLocal$ThreadLocalMap$Entry[] = {ThreadLocal$ThreadLocalMap$Entry[16]@818}
     不显示空元素
     1 = {ThreadLocal$ThreadLocalMap$Entry@820}
         value: Object = "5"
         referent: Object = {ThreadLocal@704}
         queue: ReferenceQueue = {ReferenceQueue$Null@829}
         next: Reference = null
         discovered: Reference = null
         processPendingLock: Object = {Object@827}
         processPendingActive: boolean = false
         $assertionsDisabled: boolean = true
     3 = {ThreadLocal$ThreadLocalMap$Entry@821}
         value: Object = "3"
         referent: Object = {ThreadLocal@831}
         queue: ReferenceQueue = {ReferenceQueue$Null@829}
```

图 4-29　Debug 模式查看线程局部变量数据

第 5 个除了在 table 字段中被引用外还在 THREAD_LOCAL 字段被引用。

注意：table 字段的数据类型是 Entry[]，其 key 的数据最终存储在 Reference 类的 referent 对象字段中。

修改 ThreadLocalMapTest 类，代码如下：

```java
//第 4 章/four/ThreadLocalMapTest.java
public class ThreadLocalMapTest {

    public static ThreadLocal<String> THREAD_LOCAL = null;

    public static Object obj = new Object();

    public static void main(String[] args) {
        obj = null;              //源对象所占用的内存空间将被回收

        for (int x = 1; x < 6; x++) {
            THREAD_LOCAL = new ThreadLocal<>();
            THREAD_LOCAL.set(String.valueOf(x));
        }
        Thread currentThread = Thread.currentThread();
        System.gc();             //手动调用垃圾回收器
        System.out.println(THREAD_LOCAL.get());
    }
}
```

执行结果如下：

5

再次用 Debug 模式查看线程局部变量的数据,发现 1～5 的数据还是存放在 table 字段中,但是除了 5 以外,其他 1～4 的 key 已经被置空,如图 4-30 所示。

```
v ⓕ threadLocals: ThreadLocal$ThreadLocalMap = {ThreadLocal$ThreadLocalMap@815}  ← Thread对象
    ⓕ INITIAL_CAPACITY: int = 16
  v ⓕ table: ThreadLocal$ThreadLocalMap$Entry[] = {ThreadLocal$ThreadLocalMap$Entry[16]@818}
      不显示空元素
    > ≡ 1 = {ThreadLocal$ThreadLocalMap$Entry@820}
      > ⓕ value: Object = "5"
      > ⓕ referent: Object = {ThreadLocal@704}
      > ⓕ queue: ReferenceQueue = {ReferenceQueue$Null@829}
        ⓕ next: Reference = null
        ⓕ discovered: Reference = null
      > ⓕ processPendingLock: Object = {Object@827}
        ⓕ processPendingActive: boolean = false
        ⓕ $assertionsDisabled: boolean = true
    v ≡ 3 = {ThreadLocal$ThreadLocalMap$Entry@821}
      > ⓕ value: Object = "3"
        ⓕ referent: Object = null   ← key被置空
      > ⓕ queue: ReferenceQueue = {ReferenceQueue$Null@829}
        ⓕ next: Reference = null
        ⓕ discovered: Reference = null
```

图 4-30　key 被置空

由此可以得出结论,弱引用其核心概念就是当一个对象地址只存在于 Reference 类的 referent 对象字段中被引用时,垃圾回收器工作时会将此字段置空。

注意：线程局部变量在使用中是否会造成内存泄漏问题？如果造成了内存泄漏问题,则是由什么原因引起的。

小结

通过本章的学习,相信读者对弱引用都有了自己的理解,要去把学到的知识转换为自己的理解,这样进步才会更大、更适合自己。线程局部变量在流行框架中都会被大量使用,去理解它的核心概念是走向高级架构的一个必经之路。

习题

1. 判断题

(1) 线程局部变量并发安全、线程之间隔离。(　　)

(2) 线程局部变量同一个 key 多次设置 value 将呈现覆盖效果。(　　)

(3) 共享线程局部变量的数据存储在线程对象的 threadLocals 字段中。(　　)

(4) 线程局部变量存储 key 为弱引用,垃圾回收器将对此进行特别处理。(　　)

(5) 线程局部变量存储数据，不同的 key 位运算也有可能定位到同一个数组索引。（　　）
(6) ThreadLocal 对象是官方操作线程局部变量的唯一入口。（　　）
(7) ThreadLocalMap 对象是包访问权限，在外部没有办法直接创建对象。（　　）
(8) ThreadLocalMap 对象的数据清理机制依赖于 key 等于空。（　　）

2．选择题

(1) 线程局部变量存储在线程对象的（　　）字段中。（多选）

  A．threadLocals   B．inheritableThreadLocals
  C．group   D．table

(2) 共享线程局部变量存储在线程对象的（　　）字段中。（单选）

  A．threadLocals   B．inheritableThreadLocals
  C．group   D．table

(3) 线程局部变量官方操作入口类有（　　）。（多选）

  A．Thread   B．ThreadLocalMap
  C．ThreadLocal   D．InheritableThreadLocal

(4) ThreadLocalMap 对象在（　　）方法中有数据清理的效果。（多选）

  A．set(ThreadLocal<?> k, Object v)   B．getEntry(ThreadLocal<?> k)
  C．remove(ThreadLocal<?> k)   D．expungeStaleEntries()

3．填空题

(1) 查看执行结果，并补充代码，代码如下：

```
//第 4 章/answer/OneAnswer.java
public class OneAnswer {

    public static ThreadLocal<String> THREAD_LOCAL = _____;

    public static void main(String[] args) {
        THREAD_LOCAL.set("线程局部变量");
        MyRunnable myRunnable = new MyRunnable();
        new Thread(myRunnable,_____).start();
        System.out.println(Thread.currentThread().getName()
                            + ":" + THREAD_LOCAL.get());
    }

    static final class MyRunnable implements Runnable{

        @Override
        public void run() {
            System.out.println(Thread.currentThread().getName()
                            + ":" + THREAD_LOCAL.get());
        }
    }
}
```

执行结果如下:

```
main:线程局部变量
A:null
```

(2) 查看执行结果,并补充代码,代码如下:

```java
//第 4 章/answer/TwoAnswer.java
public class TwoAnswer {

public static ThreadLocal<String> THREAD_LOCAL = _____;

    public static void main(String[] args) {
        _____;
        MyRunnable myRunnable = new MyRunnable();
        new Thread(myRunnable,"B").start();
        System.out.println(Thread.currentThread().getName()
                                    + ":" + THREAD_LOCAL.get());
    }

    static final class MyRunnable implements Runnable{

        @Override
        public void run() {
            System.out.println(Thread.currentThread().getName()
                                        + ":" + THREAD_LOCAL.get());
        }
    }
}
```

执行结果如下:

```
main:线程局部变量
B:线程局部变量
```

(3) 查看执行结果,并补充代码,代码如下:

```java
//第 4 章/answer/ThreeAnswer.java
public class ThreeAnswer {

    public static ThreadLocal<String> THREAD_LOCAL
                            = new InheritableThreadLocal<>();

    public static void main(String[] args) throws InterruptedException {
        THREAD_LOCAL.set("线程局部变量");
        MyRunnable myRunnable = new MyRunnable();
        new Thread(myRunnable,"B").start();
        Thread.sleep(1000);
        _____;
        System.out.println(Thread.currentThread().getName()
                                    + ":" + THREAD_LOCAL.get());
```

```java
    }
    static final class MyRunnable implements Runnable{

        @Override
        public void run() {
            System.out.println(Thread.currentThread().getName()
                                    + ":" + THREAD_LOCAL.get());
            try {
                Thread.sleep(2000);
            } catch (InterruptedException e) {
                throw new RuntimeException(e);
            }
            System.out.println(Thread.currentThread().getName()
                                    + ":" + THREAD_LOCAL.get());
        }
    }
}
```

执行结果如下:

```
B:线程局部变量
main:null
B:null
```

第 5 章 Lock 锁

Lock 锁在 JDK 1.5 时被推出,同期推出的并发包为 Java 多线程奠定了坚实的基础,Lock 锁提供了比 synchronized 锁更广泛的锁操作,它们允许更灵活的锁结构并且可以关联多个 Condition 对象。

Lock 锁和 synchronized 锁的对比见表 5-1。

表 5-1 Lock 锁和 synchronized 锁

锁 特 性	Lock 锁	synchronized 锁
可重入	√	√
自动释放	×	√
可中断	√	×
可尝试拿锁,有等待时间	√	×
可尝试立即拿锁	√	×
同一个锁可以有多个唤醒及等待操作对象	√	×
灵活的锁资源获取	√	×
公平锁	√	×
读写分离锁	√	×

5.1 Lock 接口

7min

Lock 接口为锁提供了标准规范的方法,子类必须实现这些方法。

1. lock()

获得锁,阻塞当前执行线程直到获得锁为止。

2. lockInterruptibly()

获得锁,阻塞当前执行线程直到获得锁为止,或者如果当前执行线程中断,则抛出 InterruptedException 异常并且清除当前执行线程的中断状态。

3. newCondition()

创建与此锁相关联的新 Condition 对象,Condition 对象可以使当前执行线程唤醒或等待,类似 Object 对象监视器功能。

4. tryLock()

尝试拿锁,获得可用的锁并立即返回 true,如果锁不可用,则立即返回 false。

5. tryLock(long time, TimeUnit unit)

尝试拿锁并最大等待给定的时间段,如果获得可用的锁,则立即返回 true,否则超过等待的时间段返回 false。接收 long 入参,作为最大等待时间;接收 TimeUnit 入参,作为时间的单位。

6. unlock()

释放锁,只有锁的持有者才能释放锁。如果当前执行线程不是锁的持有者,则会引发 IllegalMonitorStateException 异常。

5.2 ReentrantLock

ReentrantLock 实现了 Lock 接口,不仅提供了 Lock 接口标准规范的锁方法,还提供了灵活的锁资源获取。

5.2.1 构造器

ReentrantLock 构造器见表 5-2。

表 5-2 ReentrantLock 构造器

构造器	描述
ReentrantLock()	构造新的对象,默认无参构造器
ReentrantLock(boolean fair)	构造新的对象,指定是否使用公平锁

5.2.2 常用方法

1. lock()

获得锁,如果锁未被持有,则获得该锁,并将锁保持计数设置为 1;如果当前执行线程已经持有此锁,则保持计数将增加 1;如果锁由另一个执行线程持有,则当前执行线程将阻塞等待,直到获得锁为止。这里提到的保持计数就是可重入的一种标记。

标准使用示例,代码如下:

```
//第 5 章/one/OneMain.java
public class OneMain {

    public static void main(String[] args) throws InterruptedException {
        LockRunnable lockRunnable = new LockRunnable();
        Thread threadA = new Thread(lockRunnable, "A");
```

```java
            Thread threadB = new Thread(lockRunnable, "B");
            threadB.start();
            threadA.start();
        }
    static class LockRunnable implements Runnable{
        private final ReentrantLock lock = new ReentrantLock();
                        //创建 ReentrantLock 对象,默认构造器为非公平锁
        @Override
        public void run() {
            lock.lock();                    //获得锁
            try {
                System.out.println(Thread.currentThread().getName());
                Thread.sleep(2000);
            } catch (Exception e) {
                e.printStackTrace();
            } finally {
                lock.unlock();              //释放锁
            }
        }
    }
}
```

执行结果如下:

```
B
A
```

2. unlock()

释放锁,如果当前执行线程是该锁的持有者,则保持计数将递减 1,若保持计数为 0,则释放锁;如果当前执行线程不是此锁的持有者,则会引发 IllegalMonitorStateException 异常。

3. lockInterruptibly()

获得锁,但比 lock()方法多了一个响应中断,代码如下:

```java
//第 5 章/one/OneMain.java
public class OneMain {

    public static void main(String[] args) throws InterruptedException {
        LockRunnable lockRunnable = new LockRunnable();
        Thread threadA = new Thread(lockRunnable, "A");
        Thread threadB = new Thread(lockRunnable, "B");
        threadB.start();
        threadA.start();
        Thread.sleep(200);
        threadB.interrupt();                //中断
        threadA.interrupt();                //中断
    }
```

```java
static class LockRunnable implements Runnable{

    private final ReentrantLock lock = new ReentrantLock();
                        //创建 ReentrantLock 对象,默认构造器为非公平锁
    @Override
    public void run() {
        try {
            lock.lockInterruptibly();           //获得锁,此方法响应中断
            System.out.println(Thread.currentThread().getName()
                                            +":lock");
            while (true){//死循环,为了演示中断效果
            }
        } catch (Exception e) {
            e.printStackTrace();
        } finally {
            if(lock.isHeldByCurrentThread()){//查询此锁是否由当前线程持有
                System.out.println(Thread.currentThread().getName()
                                            +":unlock");
                lock.unlock();            //释放锁
            }
        }
    }
}
```

执行结果如下：

```
A:lock
java.lang.InterruptedException
    at
java.base/java.util.concurrent.locks.AbstractQueuedSynchronizer.acquireInterruptibly
(AbstractQueuedSynchronizer.java:959)
    at
java.base/java.util.concurrent.locks.ReentrantLock$Sync.lockInterruptibly(ReentrantLock.
java:161)
    at
java.base/java.util.concurrent.locks.ReentrantLock.lockInterruptibly(ReentrantLock.java:
372)
    at
cn.kungreat.book.five.one.OneMain$LockRunnable.run(OneMain.java:25)
    at
java.base/java.lang.Thread.run(Thread.java:833)
```

4. isHeldByCurrentThread()

查询锁是否由当前执行线程持有,如果是,则返回值为 true,否返回值为 false。在释放锁时应配合使用,以防止出现 IllegalMonitorStateException 异常。

5. tryLock()

如果锁未被持有,则获得该锁,将锁保持计数设置为 1 并返回 true;如果当前执行线程已经持有该锁,则保持计数将增加 1 并返回 true;如果锁由另一个执行线程持有,则该方法

将立即返回 false。

即使此锁已设置使用公平策略,如果该锁可用,则调用此方法将立即获取该锁,无论其他执行线程是否正在等待该锁。这种行为在某些情况下是有用的,尽管它破坏了公平策略。

如果想遵守此锁的公平策略,则可使用 tryLock(0,TimeUnit.SECONDS),代码如下:

```java
//第 5 章/two/TwoMain.java
public class TwoMain {

    public static void main(String[] args) {
        LockRunnable lockRunnable = new LockRunnable();
        Thread threadA = new Thread(lockRunnable, "A");
        Thread threadB = new Thread(lockRunnable, "B");
        threadA.start();
        threadB.start();
    }

    static class LockRunnable implements Runnable {

        private final ReentrantLock lock = new ReentrantLock();

        @Override
        public void run() {
            if (lock.tryLock()) {//尝试拿锁
                System.out.println(Thread.currentThread().getName()
                                            + ":getLock");
                try {
                    Thread.sleep(2000);
                } catch (InterruptedException e) {
                    e.printStackTrace();
                } finally {
                    lock.unlock(); //释放锁
                }
            } else {
                System.out.println(Thread.currentThread().getName()
                                            + ":noLock");
            }

        }
    }
}
```

执行结果如下:

```
A:getLock
B:noLock
```

6. tryLock(long timeout,TimeUnit unit)

尝试拿锁并最大等待给定的时间段,如果获得可用的锁,则立即返回 true,否则超过最大的等待时间段后返回 false。接收 long 入参，作为等待锁的最大时间；接收 TimeUnit 入参，作为时间的单位。该方法比 tryLock() 多了最大等待给定的时间、响应中断、公平策略，代码如下：

```java
//第 5 章/two/TwoMain.java
public class TwoMain {

    public static void main(String[] args) {
        LockRunnable lockRunnable = new LockRunnable();
        Thread threadA = new Thread(lockRunnable, "A");
        Thread threadB = new Thread(lockRunnable, "B");
        threadA.start();
        threadB.start();
    }

    static class LockRunnable implements Runnable {

        private final ReentrantLock lock = new ReentrantLock();

        //最大等待时间充足,两个线程都可以获得锁
        @Override
        public void run() {
            try {
                if (lock.tryLock(5,TimeUnit.SECONDS)) {
                    System.out.println(Thread.currentThread().getName()
                            + ":getLock");
                    try {
                        Thread.sleep(2000);                //睡眠 2000ms
                    } catch (InterruptedException e) {
                        e.printStackTrace();
                    } finally {
                        lock.unlock();                     //释放锁
                    }
                } else {
                    System.out.println(Thread.currentThread().getName()
                            + ":noLock");
                }
            } catch (InterruptedException e) {
                e.printStackTrace();
            }
        }
    }
}
```

执行结果如下：

```
B:getLock
A:getLock
```

7. getHoldCount()

获得当前执行线程持有此锁的计数器，可以理解为锁重入的次数。如果此锁未由当前

执行线程持有,则为 0,代码如下:

```java
//第 5 章/two/OtherMethod.java
public class OtherMethod {

    public static void main(String[] args) throws InterruptedException {
        LockRunnable lockRunnable = new LockRunnable();
        Thread threadA = new Thread(lockRunnable, "A");
        threadA.start();
        Thread.sleep(500);                       //睡眠 500ms
        System.out.println("main:" + lockRunnable.lock.getHoldCount());
                                   //如果此锁未由当前线程持有,则为 0

    }
    static class LockRunnable implements Runnable {

        private final ReentrantLock lock = new ReentrantLock();

        @Override
        public void run() {
            lock.lock();
            try {
                System.out.println("run:" + lock.getHoldCount());
                                 //获得当前线程对此锁的计数器
                addCount();
            } finally {
                lock.unlock();
            }
            System.out.println("end:" + lock.getHoldCount());
                                //获得当前线程对此锁的计数器
        }

        public void addCount(){
            lock.lock();
            try {
                Thread.sleep(2000);
                System.out.println("addCount:" + lock.getHoldCount());
                                  //获得当前线程对此锁的计数器
            } catch (InterruptedException e) {
                e.printStackTrace();
            } finally {
                lock.unlock();
            }
        }
    }
}
```

执行结果如下:

```
run:1
main:0
```

```
addCount:2
end:0
```

修改 OtherMethod 类，代码如下：

```java
//第5章/two/OtherMethod.java
public class OtherMethod {

    public static void main(String[] args) throws InterruptedException {
        LockRunnable lockRunnable = new LockRunnable();
        Thread threadA = new Thread(lockRunnable, "A");
        threadA.start();
        Thread.sleep(500);                    //睡眠 500ms
        lockRunnable.lock.lock();
        try {
            System.out.println("main:" + lockRunnable.lock.getHoldCount());
                                              //获得当前线程对此锁的计数器
        } finally {
            lockRunnable.lock.unlock();
        }
    }
    static class LockRunnable implements Runnable {

        private final ReentrantLock lock = new ReentrantLock();

        @Override
        public void run() {
            lock.lock();
            try {
                System.out.println("run:" + lock.getHoldCount());
                                              //获得当前线程对此锁的计数器
                addCount();
            } finally {
                lock.unlock();
            }
            System.out.println("end:" + lock.getHoldCount());
                                              //获得当前线程对此锁的计数器
        }

        public void addCount(){
            lock.lock();
            try {
                Thread.sleep(2000);
                System.out.println("addCount:" + lock.getHoldCount());
                                              //获得当前线程对此锁的计数器
            } catch (InterruptedException e) {
                e.printStackTrace();
            }
        }
    }
}
```

执行结果如下：

```
run:1
addCount:2
end:1
```

注意：观察上方的输出结果，并思考为什么主线程没有输出内容，而且 JVM 也没有结束。

8. getQueueLength()

返回等待获得此锁的执行线程数量的估计值，代码如下：

```java
//第 5 章/two/MethodAll.java
public class MethodAll {

    private static final ReentrantLock LOCK = new ReentrantLock();

    public static void main(String[] args) throws InterruptedException {
        LockRunnable lockRunnable = new LockRunnable();
        Thread threadA = new Thread(lockRunnable, "A");
        threadA.start();                    //启动线程
        Thread.sleep(200);                  //睡眠 200ms
        LOCK.lock();                        //拿锁
        try {
            System.out.println(Thread.currentThread().getName());
        } catch (Exception e) {
            e.printStackTrace();
        } finally {
            LOCK.unlock();                  //释放锁
        }
    }

    static class LockRunnable implements Runnable {

        @Override
        public void run() {
            LOCK.lock();                    //拿锁
            try {
                System.out.println(Thread.currentThread().getName());
                Thread.sleep(1000);         //睡眠 1000ms
                System.out.println("getQueueLength:"
                        + LOCK.getQueueLength());
                            //返回等待获得此锁的线程数量的估计值
            } catch (InterruptedException e) {
                e.printStackTrace();
            } finally {
                LOCK.unlock();              //释放锁
            }
        }
    }
}
```

执行结果如下：

```
A
getQueueLength:1
main
```

9. hasQueuedThread(Thread thread)

查询给定线程对象是否正在等待获得此锁，返回 boolean 值。接收 Thread 入参，作为给定线程对象，代码如下：

```java
//第 5 章/two/MethodAll.java
public class MethodAll {

    private static final ReentrantLock LOCK = new ReentrantLock();

    public static void main(String[] args) throws InterruptedException {
        LockRunnable lockRunnable = new LockRunnable();
        Thread threadA = new Thread(lockRunnable, "A");
        threadA.start();                    //启动线程
        Thread.sleep(200);                  //睡眠 200ms
        LOCK.lock();                        //拿锁
        try {
            System.out.println(Thread.currentThread().getName());
        } catch (Exception e) {
            e.printStackTrace();
        } finally {
            LOCK.unlock();                  //释放锁
        }
    }

    static class LockRunnable implements Runnable {

        @Override
        public void run() {
            LOCK.lock();                    //拿锁
            try {
                System.out.println(Thread.currentThread().getName());
                                            //当前执行线程对象名称
                Thread.sleep(1000);
                ThreadGroup threadGroup =
                            Thread.currentThread().getThreadGroup();
                            //当前执行线程对象所属的线程组对象
                Thread[] threads = new Thread[threadGroup.activeCount()];
                threadGroup.enumerate(threads);
                            //将组内活动线程对象复制到指定数组中
                for (int i = 0; i < threads.length; i++) {
                    if(threads[i].getName().equals("main")){
                        System.out.println("hasQueuedThread:"
                            + LOCK.hasQueuedThread(threads[i]));
                            //查询给定线程对象是否正在等待获得此锁
```

```
                }
            }
        } catch (InterruptedException e) {
            e.printStackTrace();
        } finally {
            LOCK.unlock();  //释放锁
        }
    }
}
```

执行结果如下：

```
A
hasQueuedThread:true
main
```

10. hasQueuedThreads()

查询是否有任何执行线程正在等待获得此锁，返回 boolean 值。

11. isFair()

获得此锁是否为公平锁，返回 boolean 值。

5.2.3　公平锁或非公平锁

此类的构造器接收一个可选的公平参数 boolean 值。当设置为 true 时，即公平锁，在并发拿锁的情况下倾向于等待时间最长的执行线程优先拿锁，否则此锁不保证任何特定的并发拿锁顺序。

使用公平锁的程序一般显示出比使用非公平锁的程序更低的吞吐量(通常更慢)，但获得锁的时间差异较小，保证极端情况下不会出现饥饿状态(一个等待时间较久的执行线程，一直没有获得锁)。需要注意的是，此锁的公平性并不能保证系统 CPU 线程调度的公平性。

公平锁演示代码如下：

```
//第 5 章/two/FairTest.java
public class FairTest {

    private static final ReentrantLockMy LOCK = new ReentrantLockMy(true);

    public static void main(String[] args) throws InterruptedException {
        RunnableMy runnableMy = new RunnableMy();
        for (int i = 1; i < 6; i++) {
            new Thread(runnableMy,"线程" + i).start();
            Thread.sleep(100);              //保证线程启动的顺序
        }
    }
```

```java
static class RunnableMy implements Runnable{
    @Override
    public void run() {
        //循环两次,公平锁拿锁时需要排队,非公平锁可以直接拿锁
        for (int i = 0; i < 2; i++) {
            LOCK.lock(); //拿锁
            try {
                Thread.sleep(1500);
                System.out.println(LOCK.getQueuedThreads());
            } catch (Exception e) {
                e.printStackTrace();
            } finally {
                LOCK.unlock(); //释放锁
            }
        }
    }
}

static final class ReentrantLockMy extends ReentrantLock{
    public ReentrantLockMy(boolean fair) {
        super(fair);
    }

    @Override
    protected Collection<Thread> getQueuedThreads() {
        return super.getQueuedThreads();
    }
}
```

执行结果如下:

```
[Thread[线程 5,5,main], Thread[线程 4,5,main], Thread[线程 3,5,main], Thread[线程 2,5,main]]
[Thread[线程 1,5,main], Thread[线程 5,5,main], Thread[线程 4,5,main], Thread[线程 3,5,main]]
[Thread[线程 2,5,main], Thread[线程 1,5,main], Thread[线程 5,5,main], Thread[线程 4,5,main]]
[Thread[线程 3,5,main], Thread[线程 2,5,main], Thread[线程 1,5,main], Thread[线程 5,5,main]]
[Thread[线程 4,5,main], Thread[线程 3,5,main], Thread[线程 2,5,main], Thread[线程 1,5,main]]
[Thread[线程 5,5,main], Thread[线程 4,5,main], Thread[线程 3,5,main], Thread[线程 2,5,main]]
[Thread[线程 5,5,main], Thread[线程 4,5,main], Thread[线程 3,5,main]]
[Thread[线程 5,5,main], Thread[线程 4,5,main]]
[Thread[线程 5,5,main]]
[]
```

注意:将上方代码修改为非公平锁实现,并运行主方法观察输出结果。

5.2.4 自旋锁

自旋锁是指当一个执行线程在尝试拿锁时,如果此锁已经被其他执行线程占有,则该执

行线程将循环一定的次数不断地尝试拿锁,直到获得锁或者超过一定的循环尝试次数,代码如下:

```java
//第 5 章/two/SpinLock.java
public class SpinLock {
    public static final ReentrantLock REENTRANT_LOCK = new ReentrantLock();

    public static void main(String[] args) {
        ReentrantLockSpin reentrantLockSpin = new ReentrantLockSpin();
        new Thread(reentrantLockSpin).start();
        new Thread(reentrantLockSpin).start();
        new Thread(reentrantLockSpin).start();
        new Thread(reentrantLockSpin).start();
    }

    static final class ReentrantLockSpin implements Runnable {

        @Override
        public void run() {
            //循环一定的次数拿锁,可以理解为自旋次数
            for (int i = 0; i < 100; i++) {
                if(REENTRANT_LOCK.tryLock()){//尝试拿锁
                    try {
                        System.out.println("获得锁后执行的操作");
                        return; //已经获得锁做完任务退出
                    } finally {
                        REENTRANT_LOCK.unlock();
                    }
                }
            }
        }
    }
}
```

注意:以上代码每次运行后的输出结果可能不相同,自旋锁的好处是在合理的范围内使用可以降低 CPU 上下文的切换,坏处是当使用不当时也会造成 CPU 资源的损耗、浪费。

5.3 Condition

Condition 是一个接口,Lock 锁将 Object 对象监视器(wait、notify、notifyAll)方法分解为不同的 Condition 对象,Lock 实例对象可以通过 newCondition()方法创建多个 Condition 对象,每个 Condition 对象都可以提供类似等待、唤醒的效果。如果 Lock 锁取代了 synchronized 锁语句的使用,则 Condition 对象取代了 Object 对象监视器的使用。

1. await()

使当前执行线程阻塞等待,直到它被唤醒或中断,具有释放当前锁的特性。

2. await(long time, TimeUnit unit)

使当前执行线程阻塞等待,直到它被唤醒、中断或者超过最大等待时间,如果超过最大等待时间,则返回值为 false,否则返回值为 true,具有释放当前锁的特性。接收 long 入参,作为最大等待时间;接收 TimeUnit 入参,作为时间单位。代码如下:

```java
//第 5 章/three/ConditionTest.java
public class ConditionTest {
    final Lock lock = new ReentrantLock();
    final Condition condition = lock.newCondition();

    public void testTime() {
        lock.lock();
        try {
            System.out.println(Thread.currentThread().getName());
            System.out.println(condition.await(2, TimeUnit.SECONDS));
            System.out.println(Thread.currentThread().getName() + ":end");
        } catch (InterruptedException e) {
            e.printStackTrace();
        } finally {
            lock.unlock();
        }
    }
    public void test() {
        lock.lock();
        try {
            System.out.println(System.currentTimeMillis());
                                                        //当前系统时间的毫秒数
            Thread.sleep(5000);
            System.out.println(System.currentTimeMillis());
                                                        //当前系统时间的毫秒数
        } catch (InterruptedException e) {
            e.printStackTrace();
        } finally {
            lock.unlock();
        }
    }
    public static void main(String[] args) throws InterruptedException {
        ConditionTest conditionTest = new ConditionTest();
        new Thread(new Runnable() {
            @Override
            public void run() {
                conditionTest.testTime();
            }
        }, "A").start();
        Thread.sleep(200);
        new Thread(new Runnable() {
```

```
            @Override
            public void run() {
                conditionTest.test();
            }
        }, "B").start();
    }
}
```

执行结果如下：

```
A
1668237496111
1668237501115
false
A:end
```

3. awaitNanos(long nanosTimeout)

使当前执行线程阻塞等待，直到它被唤醒、中断或者超过最大等待时间，返回纳秒数等于入参时间减去等待时间的估计值，大于 0 表示提前唤醒，小于或等于 0 表示没有剩余时间，具有释放当前锁的特性。接收 long 入参，作为最大等待时间纳秒数，代码如下：

```
//第 5 章/three/ConditionTestTwo.java
public class ConditionTestTwo {
    final Lock lock = new ReentrantLock();
    final Condition condition = lock.newCondition();

    public void testTime() {
        lock.lock();
        try {
            System.out.println(Thread.currentThread().getName());
            System.out.println(condition.awaitNanos(10000000000L)); //10s
            System.out.println(Thread.currentThread().getName() + ":end");
        } catch (InterruptedException e) {
            e.printStackTrace();
        } finally {
            lock.unlock();
        }
    }
    public void test() {
        lock.lock();
        try {
            System.out.println(System.currentTimeMillis());
                                                        //当前系统时间的毫秒数
            Thread.sleep(3000);
            System.out.println(System.currentTimeMillis());
                                                        //当前系统时间的毫秒数
            condition.signal(); //唤醒单个等待此 Condition 对象的线程
        } catch (InterruptedException e) {
            e.printStackTrace();
```

```
        } finally {
            lock.unlock();
        }
    }


    public static void main(String[] args) throws InterruptedException {
        ConditionTest conditionTest = new ConditionTest();
        new Thread(new Runnable() {
            @Override
            public void run() {
                conditionTest.testTime();
            }
        }, "A").start();
        Thread.sleep(200);
        new Thread(new Runnable() {
            @Override
            public void run() {
                conditionTest.test();
            }
        }, "B").start();
    }
}
```

执行结果如下：

```
A
1668238608253
1668238611256
6797669700
A:end
```

4．awaitUninterruptibly()

使当前执行线程阻塞等待，直到它被唤醒，具有释放当前锁的特性。

5．signal()

唤醒单个等待此 Condition 对象的执行线程。

6．signalAll()

唤醒所有等待此 Condition 对象的执行线程。

此处模拟生产和消费的实例，数据存放在一个固定大小的缓冲区中。需要可以精准控制到具体方法中执行线程的等待或唤醒，即在缓冲区空间可用时通知到指定方法中的执行线程，就需要通过两个 Condition 实例来配合实现，代码如下：

```
//第 5 章/three/ConditionTest.java
public class ConditionTest<E> {
```

```java
final Lock lock = new ReentrantLock();
final Condition notFull = lock.newCondition();
final Condition notEmpty = lock.newCondition();

final Object[] items = new Object[100]; //数据缓冲区
int putptr, takeptr, count;

public void put(E x) throws InterruptedException {
    lock.lock();
    try {
        while (count == items.length)
            notFull.await();
        items[putptr] = x;
        if (++putptr == items.length) putptr = 0;
        ++count;
        notEmpty.signal();
        System.out.println(Thread.currentThread().getName() + "put:" + x);
    } finally {
        lock.unlock();
    }
}

public E take() throws InterruptedException {
    lock.lock();
    try {
        while (count == 0)
            notEmpty.await();
        E x = (E) items[takeptr];
        if (++takeptr == items.length) takeptr = 0;
        --count;
        notFull.signal();
        return x;
    } finally {
        lock.unlock();
    }
}

public static void main(String[] args) {
    ConditionTest<Integer> conditionTest = new ConditionTest<>();
    Random random = new Random();
    new Thread(new Runnable() {
        @Override
        public void run() {
            try {
                while (true){
                    Thread.sleep(1000);
                    int i = random.nextInt(9999);
                    conditionTest.put(i);
                }
            } catch (InterruptedException e) {
                throw new RuntimeException(e);
```

```
            }
        }
    },"A").start();
    new Thread(new Runnable() {
        @Override
        public void run() {
            try {
                while (true){
                    System.out.println(Thread.currentThread().getName()
                            + ":" + conditionTest.take());
                }
            } catch (InterruptedException e) {
                throw new RuntimeException(e);
            }
        }
    },"B").start();
    }
}
```

注意：观察上方代码，并思考可以精准控制到具体方法中执行线程的等待或唤醒。

5.4 ReentrantReadWriteLock

此类对象可以创建一个读锁和一个写锁，读锁、写锁都实现了 Lock 接口。在多线程并发时读锁之间可以共享（可同时拿锁），在多线程并发时读锁和写锁之间互斥。此类适用于读多写少的场景，读锁不支持 newCondition()，默认抛出 UnsupportedOperationException 异常。

5.4.1 构造器

ReentrantReadWriteLock 构造器见表 5-3。

表 5-3　ReentrantReadWriteLock 构造器

构造器	描述
ReentrantReadWriteLock()	构造新的对象，默认为无参构造器
ReentrantReadWriteLock(boolean fair)	构造新的对象，指定是否使用公平锁

 13min

 4min

 8min

5.4.2 共享锁和互斥锁

1. 共享锁

在多线程并发时读锁之间可以共享（可同时拿锁），代码如下：

```
//第 5 章/four/ReadWriteMain.java
public class ReadWriteMain {
```

```java
    static final ReentrantReadWriteLock REENTRANT_READ_WRITE_LOCK =
                        new ReentrantReadWriteLock();
    static final ReentrantReadWriteLock.ReadLock readLock =
            REENTRANT_READ_WRITE_LOCK.readLock(); //获得读锁

    public static void main(String[] args) {
        RunnableMy runnableMy = new RunnableMy();
        Thread threadA = new Thread(runnableMy, "A");
        Thread threadB = new Thread(runnableMy, "B");
        threadA.start();
        threadB.start();
    }

    static class RunnableMy implements Runnable{

        @Override
        public void run() {
            readLock.lock();
            try {
                System.out.println(Thread.currentThread().getName()
                                + ":" + System.currentTimeMillis());
                                //输出当前执行线程名称:当前系统时间(ms)
                Thread.sleep(5000);
            } catch (Exception e) {
                e.printStackTrace();
            } finally {
                readLock.unlock();
            }
        }
    }
}
```

执行结果如下：

```
B:1668914471426
A:1668914471427
```

2. 互斥锁

在多线程并发时读锁和写锁之间互斥，代码如下：

```java
//第 5 章/four/ReadWriteMain.java
public class ReadWriteMain {
    static final ReentrantReadWriteLock REENTRANT_READ_WRITE_LOCK =
                        new ReentrantReadWriteLock();
    static final ReentrantReadWriteLock.ReadLock readLock =
                REENTRANT_READ_WRITE_LOCK.readLock();
                                            //获得读锁
    static final ReentrantReadWriteLock.WriteLock writeLock =
                REENTRANT_READ_WRITE_LOCK.writeLock();
                                            //获得写锁
```

```java
    public static void main(String[] args) {
        RunnableMy runnableMy = new RunnableMy();
        Thread threadA = new Thread(runnableMy, "A");
        Thread threadB = new Thread(new Runnable() {
            @Override
            public void run() {
                runnableMy.writeTest();
            }
        }, "B");
        threadA.start();
        threadB.start();
    }

    static class RunnableMy implements Runnable{

        @Override
        public void run() {
            readLock.lock();
            try {
                System.out.println(Thread.currentThread().getName()
                                   + ":" + System.currentTimeMillis());
                                   //输出当前执行线程名称:当前系统时间(ms)
                Thread.sleep(5000);
            } catch (Exception e) {
                e.printStackTrace();
            } finally {
                readLock.unlock();
            }
        }

        public void writeTest(){
            writeLock.lock();
            try {
                System.out.println(Thread.currentThread().getName()
                                   + ":" + System.currentTimeMillis());
                                   //输出当前执行线程名称:当前系统时间(ms)
                Thread.sleep(5000);
            } catch (Exception e) {
                e.printStackTrace();
            } finally {
                writeLock.unlock();
            }
        }
    }
}
```

执行结果如下:

A:1668914799127
B:1668914804139

5.4.3 重入特性

1. 持有写锁时可以再获得读锁

代码如下:

```java
//第5章/four/ReentryMain.java
public class ReentryMain {
    static final ReentrantReadWriteLock REENTRANT_READ_WRITE_LOCK =
                                        new ReentrantReadWriteLock();
    static final ReentrantReadWriteLock.ReadLock readLock =
                        REENTRANT_READ_WRITE_LOCK.readLock();
    static final ReentrantReadWriteLock.WriteLock writeLock =
                        REENTRANT_READ_WRITE_LOCK.writeLock();

    public static void main(String[] args) {
        RunnableMy runnableMy = new RunnableMy();
        runnableMy.writeTest();
    }

    static class RunnableMy implements Runnable{

        @Override
        public void run() {
            readLock.lock();
            try {
                System.out.println(Thread.currentThread().getName()
                                            + ":run");
                                            //输出当前执行线程名称

            } catch (Exception e) {
                e.printStackTrace();
            } finally {
                readLock.unlock();
            }
        }

        public void writeTest(){
            writeLock.lock();
            try {
                System.out.println(Thread.currentThread().getName()
                                            +":writeTest");
                                            //输出当前执行线程名称

                run();
            } catch (Exception e) {
                e.printStackTrace();
            } finally {
                writeLock.unlock();
            }
        }
    }
}
```

执行结果如下:

```
main:writeTest
main:run
```

2. 持有读锁时不可以再获得写锁

代码如下:

```java
public class ReentryMain {
    static final ReentrantReadWriteLock REENTRANT_READ_WRITE_LOCK =
                                    new ReentrantReadWriteLock();
    static final ReentrantReadWriteLock.ReadLock readLock =
                        REENTRANT_READ_WRITE_LOCK.readLock();
    static final ReentrantReadWriteLock.WriteLock writeLock =
                        REENTRANT_READ_WRITE_LOCK.writeLock();

    public static void main(String[] args) {
        RunnableMy runnableMy = new RunnableMy();
        runnableMy.run();
    }

    static class RunnableMy implements Runnable{

        @Override
        public void run() {
            readLock.lock();
            try {
                System.out.println(Thread.currentThread().getName()
                                            + ":run");
                                    //输出当前执行线程名称
                writeTest();
            } catch (Exception e) {
                e.printStackTrace();
            } finally {
                readLock.unlock();
            }
        }

        public void writeTest(){
            writeLock.lock();
            try {
                System.out.println(Thread.currentThread().getName()
                                        + ":writeTest");
                                //输出当前执行线程名称
            } catch (Exception e) {
                e.printStackTrace();
            } finally {
                writeLock.unlock();
            }
        }
    }
}
```

注意：观察上方代码结构，一定要避免在持有读锁时再次尝试获得写锁，以上代码会造成执行线程无限期阻塞。

3. 锁降级

观察官方文档示例代码，如图 5-1 所示。

```java
class CachedData {
   Object data;
   boolean cacheValid;
   final ReentrantReadWriteLock rwl = new ReentrantReadWriteLock();

   void processCachedData() {
     rwl.readLock().lock();//获得读锁
     if (!cacheValid) {
       //在获得写锁之前必须释放读锁
       rwl.readLock().unlock();//释放读锁
       rwl.writeLock().lock();//获得写锁
       try {
         if (!cacheValid) {
           data = ...;
           cacheValid = true;
         }
         //在释放写锁之前获取读锁
         rwl.readLock().lock();
       } finally {
         rwl.writeLock().unlock(); //释放写锁，如果仍保持读取，则称之为锁降级
       }
     }

     try {
       use(data);//处理数据
     } finally {
       rwl.readLock().unlock();//释放读锁
     }
   }
}
```

图 5-1 官方文档示例代码

5.4.4 常用方法

1. isWriteLockedByCurrentThread()

查询此对象写锁是否由当前执行线程持有，如果是，则返回值为 true，否则返回值为 false，代码如下：

```java
//第 5 章/four/MethodMain.java
public class MethodMain {
    static final ReentrantReadWriteLock REENTRANT_READ_WRITE_LOCK =
                        new ReentrantReadWriteLock();
    static final ReentrantReadWriteLock.WriteLock writeLock =
                        REENTRANT_READ_WRITE_LOCK.writeLock();

    public static void main(String[] args) {
```

```java
        writeLock.lock();
        try {
            System.out.println(REENTRANT_READ_WRITE_LOCK
                                    .isWriteLockedByCurrentThread());
                            //查询此对象写锁是否由当前执行线程持有
        } catch (Exception e) {
            e.printStackTrace();
        } finally {
            writeLock.unlock();
        }
    }
}
```

执行结果如下：

```
true
```

2. getQueueLength()

返回等待获得此对象读写锁的执行线程数量的估计值，代码如下：

```java
//第5章/four/MethodMain.java
public class MethodMain {
    static final ReentrantReadWriteLock REENTRANT_READ_WRITE_LOCK =
                                new ReentrantReadWriteLock();
    static final ReentrantReadWriteLock.ReadLock readLock =
                        REENTRANT_READ_WRITE_LOCK.readLock();
    static final ReentrantReadWriteLock.WriteLock writeLock =
                        REENTRANT_READ_WRITE_LOCK.writeLock();

    public static void main(String[] args) throws InterruptedException {
        RunnableMy runnableMy = new RunnableMy();
        new Thread(runnableMy,"A").start();
        new Thread(runnableMy,"B").start();
        new Thread(new Runnable() {
            @Override
            public void run() {
                runnableMy.writeTest();
            }
        }, "C").start();
        runnableMy.writeTest();
    }

    static class RunnableMy implements Runnable{
        @Override
        public void run() {
            try {
                Thread.sleep(1000);
                readLock.lock();
                System.out.println(Thread.currentThread().getName()
                                            + ":run");
```

```
            } catch (Exception e) {
                e.printStackTrace();
            } finally {
                readLock.unlock();
            }
        }

        public void writeTest(){
            writeLock.lock();
            try {
                System.out.println(Thread.currentThread().getName()
                                            + ":writeTest");
                Thread.sleep(1500);
              System.out.println(REENTRANT_READ_WRITE_LOCK
                                            .getQueueLength());
                                //返回等待获得读写锁的执行线程数量的估计值
            } catch (Exception e) {
                e.printStackTrace();
            } finally {
                writeLock.unlock();
            }
        }
    }
```

执行结果如下：

```
main:writeTest
3
C:writeTest
2
A:run
B:run
```

注意：由于多线程并发原因，所以输出结果可能不同，但是第 1 次输出的 getQueueLength() 方法一定是 3。

3. getReadHoldCount()

获得当前执行线程持有此对象读锁的计数器，可以理解为读锁重入的次数。如果读锁未由当前执行线程持有，则为 0，代码如下：

```
//第 5 章/four/MethodMainOther.java
public class MethodMainOther {
    static final ReentrantReadWriteLock REENTRANT_READ_WRITE_LOCK =
                                            new ReentrantReadWriteLock();
    static final ReentrantReadWriteLock.ReadLock readLock =
                            REENTRANT_READ_WRITE_LOCK.readLock();
```

```java
    public static void main(String[] args) throws InterruptedException {
        RunnableMy runnableMy = new RunnableMy();
        new Thread(runnableMy,"A").start();
        new Thread(runnableMy,"B").start();
    }
    static class RunnableMy implements Runnable{
        @Override
        public void run() {
            readLock.lock();
            try {
                Thread.sleep(500);
                readTest();
            } catch (Exception e) {
                e.printStackTrace();
            } finally {
                readLock.unlock();
            }
        }

        public void readTest(){
            readLock.lock();
            try {
                Thread.sleep(100);
                System.out.println(REENTRANT_READ_WRITE_LOCK
                                            .getReadHoldCount());
                                //获得当前执行线程对此对象读锁的计数器
                Thread.sleep(500);
            } catch (Exception e) {
                e.printStackTrace();
            } finally {
                readLock.unlock();
            }
        }
    }
}
```

执行结果如下：

```
2
2
```

4. getReadLockCount()

获得此对象读锁的计数器，可以理解为读锁重入的次数，代码如下：

```java
//第 5 章/four/MethodMainOther.java
public class MethodMainOther {
    static final ReentrantReadWriteLock REENTRANT_READ_WRITE_LOCK =
                                    new ReentrantReadWriteLock();
    static final ReentrantReadWriteLock.ReadLock readLock =
                        REENTRANT_READ_WRITE_LOCK.readLock();
```

```java
public static void main(String[] args) throws InterruptedException {
    RunnableMy runnableMy = new RunnableMy();
    new Thread(runnableMy,"A").start();
    new Thread(runnableMy,"B").start();
}
static class RunnableMy implements Runnable{
    @Override
    public void run() {
        readLock.lock();
        try {
            Thread.sleep(500);
            readTest();
        } catch (Exception e) {
            e.printStackTrace();
        } finally {
            readLock.unlock();
        }
    }

    public void readTest(){
        readLock.lock();
        try {
            Thread.sleep(100);
            System.out.println(REENTRANT_READ_WRITE_LOCK
                    .getReadLockCount());
            //获得此对象读锁的计数器
            Thread.sleep(500);
        } catch (Exception e) {
            e.printStackTrace();
        } finally {
            readLock.unlock();
        }
    }
}
```

执行结果如下：

```
4
4
```

5. getWriteHoldCount()

获得当前执行线程持有此对象写锁的计数器，可以理解为写锁重入的次数。如果写锁未由当前执行线程持有，则为0。

6. getWaitQueueLength(Condition condition)

返回与此对象写锁相关联的Condition条件下的执行线程等待数量的估计值。接收Condition入参，作为给定条件对象，代码如下：

```java
//第5章/four/MethodMainWait.java
public class MethodMainWait {
    static final ReentrantReadWriteLock REENTRANT_READ_WRITE_LOCK =
                                    new ReentrantReadWriteLock();
    static final ReentrantReadWriteLock.WriteLock WRITE_LOCK =
                            REENTRANT_READ_WRITE_LOCK.writeLock();
    static final Condition condition = WRITE_LOCK.newCondition();

    public static void main(String[] args) throws InterruptedException {
        RunnableMy runnableMy = new RunnableMy();
        new Thread(runnableMy,"A").start();
        new Thread(runnableMy,"B").start();
        Thread.sleep(1000);
        WRITE_LOCK.lock();
        try {
            System.out.println(REENTRANT_READ_WRITE_LOCK
                                    .getWaitQueueLength(condition));
            condition.signalAll();
        } catch (Exception e) {
            e.printStackTrace();
        } finally {
            WRITE_LOCK.unlock();
        }
    }

    static class RunnableMy implements Runnable{
        @Override
        public void run() {
            WRITE_LOCK.lock();
            try {
                condition.await();  //使当前执行线程阻塞等待
                System.out.println(Thread.currentThread().getName());
            } catch (Exception e) {
                e.printStackTrace();
            } finally {
                WRITE_LOCK.unlock();
            }
        }
    }
}
```

执行结果如下：

```
2
B
A
```

7. hasQueuedThread(Thread thread)

查询给定线程对象是否正在等待获得此对象读锁或写锁，返回 boolean 值。接收

Thread 入参,作为给定线程对象。

8. hasQueuedThreads()

查询是否有任何执行线程正在等待获得此对象读锁或写锁,返回 boolean 值。

9. hasWaiters(Condition condition)

查询与此对象写锁相关联的 Condition 条件下的执行线程是否有正在等待获得此写锁,如果有,则为 true,否则为 false。接收 Condition 入参,作为给定条件对象。

10. isFair()

获得此锁是否为公平锁,返回 boolean 值。

小结

Lock 锁提供了比 synchronized 锁更强大、更灵活的功能,但是使用起来的复杂度也比 synchronized 锁会更高一些,需要根据具体的业务场景选择合适的锁方案,并不是说哪一方一定强,哪一方一定弱。

习题

1. 判断题

(1) Lock 锁出现异常时会自动释放锁。(　　)

(2) synchronized 锁出现异常时会自动释放锁。(　　)

(3) ReentrantReadWriteLock 读写锁可以在持有读锁时再获得写锁。(　　)

(4) ReentrantReadWriteLock 读写锁可以在持有写锁时再获得读锁。(　　)

(5) 公平锁相较非公平锁往往更慢。(　　)

(6) ReentrantReadWriteLock 读写锁读锁之间多线程并发共享。(　　)

(7) ReentrantReadWriteLock 读写锁读锁、写锁之间多线程并发互斥。(　　)

(8) ReentrantLock 对象可以创建多个 Condition 对象。(　　)

(9) ReentrantReadWriteLock 读写锁读锁不支持创建 Condition 对象。(　　)

2. 选择题

(1) Lock 锁拿锁时可以立即返回的方法是(　　)。(单选)

  A. lock()   B. lockInterruptibly()

  C. tryLock()   D. unlock()

(2) Lock 锁最大等待时间拿锁的方法是(　　)。(单选)

  A. tryLock()   B. unlock()

  C. tryLock(long time, TimeUnit unit)   D. newCondition()

(3) ReentrantLock 查询此锁是否由当前执行线程持有的方法是(　　)。(单选)

  A. isLocked()   B. hasQueuedThreads()

C. isHeldByCurrentThread()　　　　D. isFair()

(4) 以下(　　)类实现了 Lock 接口。(多选)

　　A. ReentrantLock

　　B. ReentrantReadWriteLock

　　C. ReentrantReadWriteLock.ReadLock

　　D. ReentrantReadWriteLock.WriteLock

(5) Lock 锁在公平模式下，以下(　　)方法不支持公平策略。(单选)

　　A. lock()

　　B. lockInterruptibly()

　　C. tryLock()

　　D. tryLock(long time，TimeUnit unit)

3. 填空题

(1) 根据业务要求补全代码，多线程并发拿锁时要求不阻塞并立即返回，然后输出线程名称，代码如下：

```java
//第5章/answer/TryLock.java
public class TryLock {

    private static final Lock LOCK = _____;

    public static void main(String[] args) {
        MyRunnable myRunnable = new MyRunnable();
        new Thread(myRunnable,_____).start();
        new Thread(myRunnable,_____).start();
        new Thread(myRunnable,_____).start();
        new Thread(myRunnable,_____).start();
    }

    private static final class MyRunnable implements Runnable{

        @Override
        public void run() {
            if(_____){
                try {
                    System.out.println(_____ +":获得锁");
                } finally {
                    LOCK.unlock();
                }
            }else{
                System.out.println(_____ +":没获得锁");
            }
        }
    }
}
```

(2) 根据业务要求补全代码，参考官方文档完成锁降级，代码如下：

```java
//第 5 章/answer/CachedData.java
public class CachedData {
    static final ReentrantReadWriteLock rwl = new ReentrantReadWriteLock();
    static boolean cacheValid = false;

    public static void main(String[] args) {
        MyRunnable myRunnable = new MyRunnable();
        new Thread(myRunnable,"A").start();
        new Thread(myRunnable,"B").start();
        new Thread(myRunnable,"C").start();
        new Thread(myRunnable,"D").start();
    }

    private static final class MyRunnable implements Runnable{

        @Override
        public void run() {
            rwl.readLock().lock();
            if (!cacheValid) {
                _____;
                rwl.writeLock().lock();
                try {
                    if (!cacheValid) {
                        System.out.println("模拟数据写操作...");
                        cacheValid = true;
                    }
                    //通过在释放写锁之前获得读锁来完成锁降级
                    rwl.readLock().lock();
                } finally {
                    _____;
                }
            }
            try {
                System.out.println("模拟数据读操作...");
            } finally {
                rwl.readLock().unlock();
            }
        }
    }
}
```

(3) 根据业务要求补全代码,参考官方文档完成读写锁,代码如下:

```java
//第 5 章/answer/RWDictionary.java
public class RWDictionary {
    private final Map<String, Object> m = new TreeMap<>();
    private final ReentrantReadWriteLock rwl = new ReentrantReadWriteLock();
    private final Lock r = rwl.readLock();
    private final Lock w = rwl.writeLock();
```

```
    public Object get(String key) {
        _____;
        try {
            return m.get(key);
        } finally {
            _____;
        }
    }

    public List<String> allKeys() {
        r.lock();
        try {
            return new ArrayList<>(m.keySet());
        } finally {
            r.unlock();
        }
    }

    public Object put(String key, Object value) {
        _____;
        try {
            return m.put(key, value);
        } finally {
            _____;
        }
    }

    public void clear() {
        _____
        try {
            m.clear();
        } finally {
            _____;
        }
    }
}
```

（4）根据业务要求补全代码，A、B 两个执行线程将按照 1～100 的顺序轮流打印数字，代码如下：

```
//第 5 章/answer/LoopPrint.java
public class LoopPrint {
    private static final ReentrantLock REENTRANT_LOCK = new ReentrantLock();
    private static final Condition CONDITION = REENTRANT_LOCK.newCondition();

    public static void main(String[] args) throws InterruptedException {
        LoopRunnable loopRunnable = new LoopRunnable();
        new Thread(loopRunnable, "A").start();
        new Thread(loopRunnable, "B").start();
    }
```

```java
    static class LoopRunnable implements Runnable {
        private volatile int num = 1;

        @Override
        public void run() {
            REENTRANT_LOCK.lock();
            try {
                for (; num <= 100; ) {
                    System.out.println(Thread.currentThread().getName()
                            + ":" + num);
                    num++;
                    _____;
                    _____;
                }
            } catch (InterruptedException e) {
                e.printStackTrace();
            } finally {
                _____;
                REENTRANT_LOCK.unlock();
            }
        }
    }
}
```

（5）根据业务要求补全代码，启动 3 个执行线程 A、B、C，每个执行线程将自己的名称轮流打印 5 遍，打印顺序是 ABCABC…，代码如下：

```java
//第 5 章/answer/ThreePrint.java
public class ThreePrint {

    private static final ReentrantLock LOCK = new ReentrantLock();
    private static final Condition CONDITION = _____;

    public static void main(String[] args) {
        LoopRunnable loopRunnable = new LoopRunnable();
        new Thread(loopRunnable, "A").start();
        new Thread(loopRunnable, "B").start();
        new Thread(loopRunnable, "C").start();
    }

    static class LoopRunnable implements Runnable {
        private volatile int loopIndex = 0;
        private final String[] loopNames = {"A", "B", "C"};

        @Override
        public void run() {
            for (int x = 0; x < 5; x++) {
                LOCK.lock();
                try {
                    String name = Thread.currentThread().getName();
```

```
                //名称不匹配时一直循环
                while (!name.equals(loopNames[loopIndex])) {
                    _____;
                }
                System.out.print(name);             //消费名称
                loopIndex++;
                if (loopIndex == loopNames.length) {
                    loopIndex = 0;                  //重置指针
                }
                _____;
            } catch (InterruptedException e) {
                e.printStackTrace();
            } finally {
                LOCK.unlock();
            }
        }
    }
}
```

(6) 模拟从 1 累加到 100，启动 10 个执行线程平分此任务，主线程需要等待其他执行线程任务完成，然后统计计算结果并输出，代码如下：

```
//第 5 章/answer/TenCount.java
public class TenCount {

    public static void main(String[] args) {
        FutureRun[] futureRuns = new FutureRun[10];
        for (int i = 0; i < futureRuns.length; i++) {
            FutureRun futureRun = new FutureRun(i * 10 + 1);
            futureRuns[i] = futureRun;
            new Thread(futureRun).start();
        }
        int numCount = 0;
        for (int i = 0; i < futureRuns.length; i++) {
            numCount = numCount + futureRuns[i].getCountNum();
        }
        System.out.println("总结果:" + numCount);
    }

    static final class FutureRun implements Runnable {

        private final ReentrantLock reentrantLock = new ReentrantLock();
        private final Condition condition = reentrantLock.newCondition();
        private final int startNum;
        private boolean isEnd = false;
        private int countNum;

        public FutureRun(int startNum) {
            this.startNum = startNum;
```

```java
    }
    @Override
    public void run() {
        int endNum = startNum + 10;
        reentrantLock.lock();
        try {
            for (int i = startNum; i < endNum; i++) {
                countNum = countNum + i;
                Thread.sleep(1000);
            }
            System.out.println(Thread.currentThread().getName()
                                                    + ":done");
        } catch (InterruptedException e) {
            countNum = 0;
            e.printStackTrace();
        } finally {
            isEnd = true;
            _____;
            reentrantLock.unlock();
        }
    }

    public int getCountNum() {
        reentrantLock.lock();
        try {
            while (!isEnd) {
                _____;
            }
            return countNum;
        } catch (InterruptedException e) {
            e.printStackTrace();
        } finally {
            reentrantLock.unlock();
        }
        return 0;
    }
}
```

注意：此案例都在演示执行线程之间如何协作，FutureRun 对象的数据并没有多线程并发操作。

第 6 章 atomic 原子包

原子性指事务的不可分割性,一个事务内的所有操作要么全部执行成功,要么相当于没有执行。原子操作由 CPU 和底层系统设计实现,具有多线程并发互斥效果,比锁的粒度更小,通常用于修改某个具体的值。

6.1 AtomicBoolean

以原子方式更新 boolean 值,真实存储的数据是 int 值,中间会有 int 值转换为 boolean 值的处理,底层实现基于 VarHandle(JDK 9 新增的类)实现原子操作,如图 6-1 所示。

15min

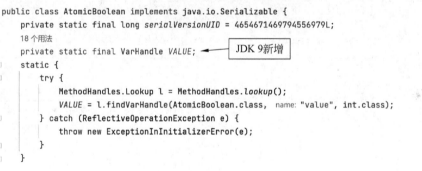

6min

9min

图 6-1 AtomicBoolean 源代码

6.1.1 构造器

AtomicBoolean 构造器见表 6-1。

表 6-1 AtomicBoolean 构造器

构 造 器	描 述
AtomicBoolean()	构造新的对象,默认为无参构造器
AtomicBoolean(boolean initialValue)	构造新的对象,指定初始值

6.1.2 常用方法

1. setPlain(boolean newValue)

设置数据,其内存语义就像字段被声明为非 volatile 修饰的效果一样。接收 boolean 入参,作为新的数据,代码如下:

```java
//第 6 章/one/AtomicMain.java
public class AtomicMain {

    private static final AtomicBoolean ATOMIC_BOOLEAN = new AtomicBoolean();

    public static void main(String[] args) throws InterruptedException {
        AtomicRun atomicRun = new AtomicRun();
        new Thread(atomicRun,"A").start();
        Thread.sleep(1000);
        atomicRun.setPlain();
    }

    static final class AtomicRun implements Runnable{

        @Override
        public void run() {
            while (!ATOMIC_BOOLEAN.getPlain()){
//返回当前值,其读取的内存语义就像字段被声明为非 volatile 修饰的效果一样
            }
            System.out.println(Thread.currentThread().getName() + ":end");
        }

        public void setPlain(){
            ATOMIC_BOOLEAN.setPlain(true);
            //设置数据,其内存语义就像字段被声明为非 volatile 修饰的效果一样
        }
    }
}
```

注意: 回顾 volatile 关键字解决的问题的可见性、重排序。运行上述代码后会产生死循环,这是由于可见性问题导致的。

2. getPlain()

返回当前值,其读取的内存语义就像字段被声明为非 volatile 修饰的效果一样。

3. set(boolean newValue)

设置数据,其内存语义就像字段被声明为 volatile 修饰的效果一样。接收 boolean 入参,作为新的数据,代码如下:

```java
//第 6 章/one/AtomicMain.java
public class AtomicMain {

    private static final AtomicBoolean ATOMIC_BOOLEAN = new AtomicBoolean();

    public static void main(String[] args) throws InterruptedException {
        AtomicRun atomicRun = new AtomicRun();
        new Thread(atomicRun,"A").start();
        Thread.sleep(1000);
        atomicRun.set();
    }

    static final class AtomicRun implements Runnable{

        @Override
        public void run() {
            while (!ATOMIC_BOOLEAN.get()){
            //返回当前值,其读取的内存语义就像字段被声明为 volatile 修饰的效果一样
            }
            System.out.println(Thread.currentThread().getName() + ":end");
        }

        public void set(){
            ATOMIC_BOOLEAN.set(true);
            //设置数据,其内存语义就像字段被声明为 volatile 修饰的效果一样
        }

    }
}
```

执行结果如下：

```
A:end
```

观察源代码,会发现此方法直接修改了 int 数据,而 int 数据是由 volatile 修饰的,如图 6-2 所示。

注意：推荐使用此方法设置新数据,此方法可以保证解决可见性、重排序问题。

4. get()

返回当前值,其读取的内存语义就像字段被声明为 volatile 修饰的效果一样,中间会有 int 值转换为 boolean 值的处理,如图 6-3 所示。

```
public final void set(boolean newValue) {
    value = newValue ? 1 : 0;
}
```

图 6-2　set(boolean newValue)源代码

```
public final boolean get() {
    return value != 0;
}
```

图 6-3　get()源代码

5. setOpaque(boolean newValue)

设置数据,其内存语义就像 VarHandle.setOpaque(java.lang.Object…)效果一样,可以解决可见性问题,但不保证重排序效果。接收 boolean 入参,作为新的数据,代码如下:

```java
//第 6 章/one/AtomicMain.java
public class AtomicMain {

    private static final AtomicBoolean ATOMIC_BOOLEAN = new AtomicBoolean();

    public static void main(String[] args) throws InterruptedException {
        AtomicRun atomicRun = new AtomicRun();
        new Thread(atomicRun,"A").start();
        Thread.sleep(1000);
        atomicRun.setOpaque();
    }

    static final class AtomicRun implements Runnable{

        @Override
        public void run() {
            while (!ATOMIC_BOOLEAN.getOpaque()){

            }
            System.out.println(Thread.currentThread().getName() + ":end");
        }

        public void setOpaque(){
            ATOMIC_BOOLEAN.setOpaque(true);
        }

    }
}
```

执行结果如下:

```
A:end
```

6. getOpaque()

返回当前值,其读取的内存语义就像 VarHandle.getOpaque(java.lang.Object…)效果一样,可以解决可见性问题,但不保证重排序效果。

7. setRelease(boolean newValue)

设置数据,其内存语义就像 VarHandle.setRelease(java.lang.Object…)效果一样,可以解决可见性问题并确保在此访问之后不会对先前的加载和存储进行重新排序。接收 boolean 入参,作为新的数据,代码如下:

```java
//第 6 章/one/AtomicMain.java
public class AtomicMain {
```

```java
    private static final AtomicBoolean ATOMIC_BOOLEAN = new AtomicBoolean();
    public static void main(String[] args) throws InterruptedException {
        AtomicRun atomicRun = new AtomicRun();
        new Thread(atomicRun,"A").start();
        Thread.sleep(1000);
        atomicRun.setRelease();
    }

    static final class AtomicRun implements Runnable{

        @Override
        public void run() {
            while (!ATOMIC_BOOLEAN.getAcquire()){

            }
            System.out.println(Thread.currentThread().getName() + ":end");
        }

        public void setRelease(){
            ATOMIC_BOOLEAN.setRelease(true);
        }

    }
}
```

执行结果如下：

```
A:end
```

注意：以上方法说明由官方文档机器翻译过来，由于重排序的问题难以验证，所以不必纠结于此，并不推荐使用此方法。

8. getAcquire()

返回当前值，其读取的内存语义就像 VarHandle.getAcquire(java.lang.Object…)效果一样，可以解决可见性问题并确保在此访问之后不会对先前的加载和存储进行重新排序。

9. compareAndSet(boolean expectedValue, boolean newValue)

如果当前值等于预期值，则更新数据并返回 true，否则返回 false。其内存语义就像 VarHandle.compareAndSet(java.lang.Object…)效果一样。分别接收 boolean 入参，作为预期值；接收 boolean 入参，作为新的数据。此方法类似于乐观锁概念，代码如下：

```java
//第 6 章/one/AtomicMain.java
public class AtomicMain {

    private static final AtomicBoolean ATOMIC_BOOLEAN = new AtomicBoolean();
```

```java
    public static void main(String[] args) throws InterruptedException {
        System.out.println(ATOMIC_BOOLEAN.get());
        System.out.println("cas:"
                + ATOMIC_BOOLEAN.compareAndSet(false, true));
        System.out.println(ATOMIC_BOOLEAN.get());
    }
}
```

执行结果如下:

```
false
cas:true
true
```

注意:CAS 操作推荐使用此方法,相当于字段有 volatile 修饰效果。

10. compareAndExchange(boolean expectedValue, boolean newValue)

如果当前值等于预期值,则更新数据并返回预期值,否则返回 false。其内存语义就像 VarHandle.compareAndExchange (java.lang.Object…)效果一样。分别接收 boolean 入参,作为预期值;接收 boolean 入参,作为新的数据。此方法类似于乐观锁概念,代码如下:

```java
//第 6 章/one/AtomicMain.java
public class AtomicMain {

    private static final AtomicBoolean ATOMIC_BOOLEAN = new AtomicBoolean();

    public static void main(String[] args) throws InterruptedException {
        System.out.println(ATOMIC_BOOLEAN.get());
        System.out.println("cas:"
                + ATOMIC_BOOLEAN.compareAndExchange(false, true));
        System.out.println(ATOMIC_BOOLEAN.get());
    }
}
```

执行结果如下:

```
false
cas:false
true
```

注意:此方法相当于字段有 volatile 修饰效果,返回值有些绕,故不建议使用。

6.2 AtomicInteger

以原子方式更新 int 值,底层基于 Unsafe 类实现原子操作,如图 6-4 所示。

27min

```
public class AtomicInteger extends Number implements java.io.Serializable {
    private static final long serialVersionUID = 6214790243416807050L;

    /*
     * 该类打算使用VarHandles实现,但是未解析的循环启动依赖项
     *
     */
    private static final Unsafe U = Unsafe.getUnsafe();
    private static final long VALUE
        = U.objectFieldOffset(AtomicInteger.class, "value");

    private volatile int value;    ←── 数据存储字段
```

图 6-4 AtomicInteger 源代码

6.2.1 构造器

AtomicInteger 构造器见表 6-2。

表 6-2 AtomicInteger 构造器

构 造 器	描 述
AtomicInteger()	构造新的对象,默认为无参构造器
AtomicInteger(int initialValue)	构造新的对象,指定初始值

6.2.2 常用方法

1. setPlain(int newValue)

设置数据,其内存语义就像字段被声明为非 volatile 修饰的效果一样。接收 int 入参,作为新的数据。

2. getPlain()

返回当前值,其读取的内存语义就像字段被声明为非 volatile 修饰的效果一样。

3. set(int newValue)

设置数据,其内存语义就像字段被声明为 volatile 修饰的效果一样。接收 int 入参,作为新的数据。

4. get()

返回当前值,其读取的内存语义就像字段被声明为 volatile 修饰的效果一样。

5. setOpaque(int newValue)

设置数据,其内存语义就像 VarHandle.setOpaque(java.lang.Object…)效果一样,可

以解决可见性问题,但不保证重排序效果。接收 int 入参,作为新的数据。

6. getOpaque()

返回当前值,其读取的内存语义就像 VarHandle.getOpaque(java.lang.Object…)效果一样,可以解决可见性问题,但不保证重排序效果。

7. setRelease(int newValue)

设置数据,其内存语义就像 VarHandle.setRelease(java.lang.Object…)效果一样,可以解决可见性问题并确保在此访问之后不会对先前的加载和存储进行重新排序。接收 int 入参,作为新的数据。

8. getAcquire()

返回当前值,其读取的内存语义就像 VarHandle.getAcquire(java.lang.Object…)效果一样,可以解决可见性问题并确保在此访问之后不会对先前的加载和存储进行重新排序。

9. compareAndSet(int expectedValue, int newValue)

如果当前值等于预期值,则更新数据并返回 true,否则返回 false。其内存语义就像 VarHandle.compareAndSet(java.lang.Object…)效果一样。分别接收 int 入参,作为预期值;接收 int 入参,作为新的数据。

10. accumulateAndGet(int x, IntBinaryOperator accumulatorFunction)

设置数据,其内存语义就像 VarHandle.compareAndSet(java.lang.Object…)效果一样,回调给定函数传入当前值和方法接收的值,然后把函数的返回结果作为更新的值。接收 int 入参,作为方法接收的值;接收 IntBinaryOperator 入参,作为给定函数对象,代码如下:

```
//第 6 章/two/AtomicInt.java
public class AtomicInt {

    public static final AtomicInteger ATOMIC_INTEGER = new AtomicInteger(10);

    public static void main(String[] args) throws Exception {
        System.out.println(ATOMIC_INTEGER.accumulateAndGet(10,
                                        (left, right) -> left + right));
    }

}
```

执行结果如下:

```
20
```

观察此方法源代码,此方法使用自旋锁更新数据,直到成功,如图 6-5 所示。

11. addAndGet(int delta)

先增加数据再返回,数据等于当前数据加上传入数据。其内存语义就像 VarHandle.getAndAdd(java.lang.Object…)效果一样,接收 int 入参,作为传入数据,代码如下:

```java
    public final int accumulateAndGet(int x,
                                      IntBinaryOperator accumulatorFunction) {
        int prev = get(), next = 0;
        for (boolean haveNext = false;;) {   // ← 自旋
            if (!haveNext)
                next = accumulatorFunction.applyAsInt(prev, x);
            if (weakCompareAndSetVolatile(prev, next))
                return next;                  // ← 更新成功才返回
            haveNext = (prev == (prev = get()));
        }
    }
```

图 6-5　accumulateAndGet 方法源代码

```java
//第 6 章/two/AtomicInt.java
public class AtomicInt {

    public static final AtomicInteger ATOMIC_INTEGER = new AtomicInteger(10);

    public static void main(String[] args) throws Exception {
        System.out.println(ATOMIC_INTEGER.addAndGet(20)); //30
    }
}
```

执行结果如下：

```
30
```

12. decrementAndGet()

先递减当前值，然后返回。其内存语义就像 VarHandle.getAndAdd(java.lang.Object…) 效果一样，代码如下：

```java
//第 6 章/two/AtomicInt.java
public class AtomicInt {

    public static final AtomicInteger ATOMIC_INTEGER = new AtomicInteger(10);

    public static void main(String[] args) throws Exception {
        System.out.println(ATOMIC_INTEGER.decrementAndGet());
    }
}
```

执行结果如下：

```
9
```

13. getAndDecrement()

先返回当前值，然后递减，其内存语义就像 VarHandle.getAndAdd(java.lang.Object…)效

果一样,代码如下:

```java
//第 6 章/two/AtomicInt.java
public class AtomicInt {

    public static final AtomicInteger ATOMIC_INTEGER = new AtomicInteger(10);

    public static void main(String[] args) throws Exception {
        System.out.println(ATOMIC_INTEGER.getAndDecrement());
        System.out.println(ATOMIC_INTEGER.get());
    }
}
```

执行结果如下:

```
10
9
```

14. incrementAndGet()

先递增当前值,然后返回。其内存语义就像 VarHandle.getAndAdd(java.lang.Object…)效果一样,代码如下:

```java
//第 6 章/two/AtomicCount.java
public class AtomicCount {
    public static final AtomicInteger ATOMIC_INTEGER_ADD =
                                                    new AtomicInteger();
    public static volatile int count;

    public static void main(String[] args) throws Exception {
        RunIncrement runIncrement = new RunIncrement();
        Thread thread = new Thread(runIncrement);
        Thread thread1 = new Thread(runIncrement);
        thread.start();
        thread1.start();

        thread.join();
        thread1.join();
        System.out.println("原子:" + ATOMIC_INTEGER_ADD.get());
        System.out.println("count:" + count);
    }

    static final class RunIncrement implements Runnable{

        @Override
        public void run() {
            for (int i = 0; i < 100000; i++) {
```

```
                count++;
                ATOMIC_INTEGER_ADD.incrementAndGet();
            }
        }
    }
}
```

执行结果如下：

```
原子:200000
count:169115
```

注意：观察结果会发现count的值每次都可能不一样，这是由于并发脏读导致的。读者应理解原子操作的意义。

6.3 AtomicReference

以原子方式更新对象引用，底层基于VarHandle（JDK 9 新增的类）实现原子操作，如图 6-6 所示。

```java
public class AtomicReference<V> implements java.io.Serializable {
    private static final long serialVersionUID = -1848883965231344442L;
    private static final VarHandle VALUE;
    static {
        try {
            MethodHandles.Lookup l = MethodHandles.lookup();
            VALUE = l.findVarHandle(AtomicReference.class, "value", Object.class);
        } catch (ReflectiveOperationException e) {
            throw new ExceptionInInitializerError(e);
        }
    }

    @SuppressWarnings("serial")
    private volatile V value;    ← 泛型数据

    public AtomicReference(V initialValue) {
        value = initialValue;
    }

    public AtomicReference() {
    }
```

图 6-6　AtomicReference 源代码

6.3.1 构造器

AtomicReference 构造器见表 6-3。

表 6-3 AtomicReference 构造器

构造器	描述
AtomicReference()	构造新的对象,默认为无参构造器
AtomicReference(V initialValue)	构造新的对象,指定初始值

6.3.2 常用方法

1. setPlain(V newValue)

设置数据,其内存语义就像字段被声明为非 volatile 修饰的效果一样。接收泛型入参,作为新的数据。

2. getPlain()

返回当前值,其读取的内存语义就像字段被声明为非 volatile 修饰的效果一样。

3. set(V newValue)

设置数据,其内存语义就像字段被声明为 volatile 修饰的效果一样。接收泛型入参,作为新的数据。

4. get()

返回当前值,其读取的内存语义就像字段被声明为 volatile 修饰的效果一样。

5. setOpaque(V newValue)

设置数据,其内存语义就像 VarHandle.setOpaque(java.lang.Object…)效果一样,可以解决可见性问题,但不保证重排序效果。接收泛型入参,作为新的数据。

6. getOpaque()

返回当前值,其读取的内存语义就像 VarHandle.getOpaque(java.lang.Object…)效果一样,可以解决可见性问题,但不保证重排序效果。

7. setRelease(V newValue)

设置数据,其内存语义就像 VarHandle.setRelease(java.lang.Object…)效果一样,可以解决可见性问题并确保在此访问之后不会对先前的加载和存储进行重新排序。接收泛型入参,作为新的数据。

8. getAcquire()

返回当前值,其读取的内存语义就像 VarHandle.getAcquire(java.lang.Object…)效果一样,可以解决可见性问题并确保在此访问之后不会对先前的加载和存储进行重新排序。

9. compareAndSet(V expectedValue, V newValue)

如果当前值等于预期值,则更新数据并返回 true,否则返回 false。其内存语义就像 VarHandle.compareAndSet(java.lang.Object…)效果一样。分别接收泛型入参,作为预期值;接收泛型入参,作为新的数据,代码如下:

```java
//第6章/three/AtomicReferenceTest.java
public class AtomicReferenceTest {

    private static final AtomicReference<User> ATOMIC_REFERENCE =
                                    new AtomicReference<>();

    public static void main(String[] args) {
        User user1 = new User("kungreat",18);
        User user2 = new User("xiangyangit",28);
        User user3 = new User("java",29);
        System.out.println(ATOMIC_REFERENCE.compareAndSet(null, user1));
        System.out.println(ATOMIC_REFERENCE.get());
        System.out.println(ATOMIC_REFERENCE.compareAndSet(null, user2));
    }

    static final class User{
        private String name;
        private Integer age;

        public User(String name, Integer age) {
            this.name = name;
            this.age = age;
        }

        public String getName() {
            return name;
        }

        public void setName(String name) {
            this.name = name;
        }

        public Integer getAge() {
            return age;
        }

        public void setAge(Integer age) {
            this.age = age;
        }

        @Override
        public String toString() {
            return "User{" +
                    "name='" + name + '\'' +
                    ", age=" + age +
                    '}';
        }
    }
}
```

执行结果如下：

```
true
User{name = 'kungreat', age = 18}
False
```

10. accumulateAndGet(V x,BinaryOperator < V > accumulatorFunction)

设置数据,其内存语义就像 VarHandle.compareAndSet(java.lang.Object…)效果一样,回调给定函数传入当前值和方法接收的值,然后把函数返回结果作为更新的值。接收泛型入参,作为方法接收的值;接收 BinaryOperator 入参,作为给定函数对象,代码如下:

```java
//第 6 章/three/AtomicReferenceTest.java
public class AtomicReferenceTest {

    private static final AtomicReference < User > ATOMIC_REFERENCE =
                                        new AtomicReference <>();

    public static void main(String[] args) {
        User user1 = new User("kungreat",18);
        System.out.println(ATOMIC_REFERENCE.accumulateAndGet(user1,
        new BinaryOperator < User >() {
            @Override
            public User apply(User old, User nw) {
                System.out.println(old); //初始值 null
                return nw;
            }
        }));
    }

    static final class User{
        private String name;
        private Integer age;

        public User(String name, Integer age) {
            this.name = name;
            this.age = age;
        }

        public String getName() {
            return name;
        }

        public void setName(String name) {
            this.name = name;
        }

        public Integer getAge() {
            return age;
        }

        public void setAge(Integer age) {
            this.age = age;
        }
```

```
        @Override
        public String toString() {
            return "User{" +
                    "name = '" + name + '\'' +
                    ", age = " + age +
                    '}';
        }
    }
}
```

执行结果如下：

```
null
User{name = 'kungreat', age = 18}
```

观察此方法源代码，此方法使用自旋锁更新数据，直到成功，如图 6-7 所示。

```
public final V accumulateAndGet(V x,
                                BinaryOperator<V> accumulatorFunction) {
    V prev = get(), next = null;
    for (boolean haveNext = false;;) {    ← 自旋锁
        if (!haveNext)
            next = accumulatorFunction.apply(prev, x);
        if (weakCompareAndSetVolatile(prev, next))
            return next;
        haveNext = (prev == (prev = get()));
    }
}
```

图 6-7　accumulateAndGet 方法源代码

6.4　AtomicIntegerFieldUpdater

一种基于反射的实用程序，支持对指定对象的 volatile 修饰的 int 字段进行原子更新。底层基于 Unsafe 实现原子操作。

6.4.1　对象创建

需要通过 newUpdater(Class< U > tclass, String fieldName)静态方法来创建此对象，如图 6-8 所示。

```
@CallerSensitive
public static <U> AtomicIntegerFieldUpdater<U> newUpdater(Class<U> tclass,
                                                          String fieldName) {
    return new AtomicIntegerFieldUpdaterImpl<U>
        (tclass, fieldName, Reflection.getCallerClass());
}
```

图 6-8　newUpdater 方法源代码

6.4.2 常用方法

1. get(T obj)

返回当前值,其读取的内存语义就像字段被声明为 volatile 修饰的效果一样。接收泛型入参,作为包装对象,代码如下:

```java
//第6章/four/AtomicIntegerFieldUpdaterTest.java
public class AtomicIntegerFieldUpdaterTest {

    private static final AtomicIntegerFieldUpdater<BaseInt> FIELD_UPDATER =
            AtomicIntegerFieldUpdater.newUpdater(BaseInt.class,"num");

    public static void main(String[] args) {
        BaseInt baseInt = new BaseInt();
        FIELD_UPDATER.set(baseInt,100);
        System.out.println(FIELD_UPDATER.get(baseInt));

    }

    static final class BaseInt {
        public volatile int num;
    }
}
```

执行结果如下:

```
100
```

2. set(T obj, int newValue)

设置数据,其内存语义就像字段被声明为 volatile 修饰的效果一样。接收泛型入参,作为包装对象;接收 int 入参,作为新的数据。

3. compareAndSet(T obj, int expect, int update)

如果当前值等于预期值,则更新数据并返回 true,否则返回 false。分别接收泛型入参,作为包装对象;接收 int 入参,作为预期值;接收 int 入参,作为新的数据,代码如下:

```java
//第6章/four/AtomicIntegerFieldUpdaterTest.java
public class AtomicIntegerFieldUpdaterTest {

    private static final AtomicIntegerFieldUpdater<BaseInt> FIELD_UPDATER =
            AtomicIntegerFieldUpdater.newUpdater(BaseInt.class,"num");

    public static void main(String[] args) {
        BaseInt baseInt = new BaseInt();
        System.out.println(FIELD_UPDATER.compareAndSet(baseInt,0,100));
        System.out.println(FIELD_UPDATER.get(baseInt));
    }
```

```
static final class BaseInt {
    public volatile int num;
}
```

执行结果如下：

```
true
100
```

6.5　AtomicIntegerArray

一个 int 数组,其中的元素可以实现原子更新。底层基于 VarHandle(JDK 9 新增的类)实现原子操作,如图 6-9 所示。

```
public class AtomicIntegerArray implements java.io.Serializable {
    private static final long serialVersionUID = 2862133569453604235L;
    private static final VarHandle AA
        = MethodHandles.arrayElementVarHandle(int[].class);
    private final int[] array;    ← 数据存储字段

    public AtomicIntegerArray(int length) {
        array = new int[length];
    }

    public AtomicIntegerArray(int[] array) {
        // Visibility guaranteed by final field guarantees
        this.array = array.clone();
    }
}
```

图 6-9　AtomicIntegerArray 源代码

6.5.1　构造器

AtomicIntegerArray 构造器见表 6-4。

表 6-4　AtomicIntegerArray 构造器

构造器	描述
AtomicIntegerArray(int length)	构造新的对象,指定数组长度
AtomicIntegerArray(int[] array)	构造新的对象,指定初始数组

6.5.2　常用方法

1. set(int i, int newValue)

设置数据,其内存语义就像字段被声明为 volatile 修饰的效果一样。分别接收 int 入参,作为数组索引;接收 int 入参,作为新的数据,代码如下：

```java
//第6章/five/AtomicArrayTest.java
public class AtomicArrayTest {

    private static final AtomicIntegerArray ATOMIC_INTEGER_ARRAY = new AtomicIntegerArray(16);

    public static void main(String[] args) {
        ATOMIC_INTEGER_ARRAY.set(6,100);
        System.out.println(ATOMIC_INTEGER_ARRAY.get(6));
        System.out.println(ATOMIC_INTEGER_ARRAY.get(15));
    }
}
```

执行结果如下：

```
100
0
```

2. get(int i)

返回当前值，其读取的内存语义就像字段被声明为 volatile 修饰的效果一样。接收 int 入参，作为数组索引。

3. compareAndSet(int i, int expectedValue, int newValue)

如果当前值等于预期值，则更新数据并返回 true，否则返回 false。其内存语义就像 VarHandle.compareAndSet(java.lang.Object…)效果一样。分别接收 int 入参，作为数组索引；接收 int 入参，作为预期值；接收 int 入参，作为新的数据，代码如下：

```java
//第6章/five/AtomicArrayTest.java
public class AtomicArrayTest {

    private static final AtomicIntegerArray ATOMIC_INTEGER_ARRAY = new AtomicIntegerArray(16);

    public static void main(String[] args) {
        ATOMIC_INTEGER_ARRAY.set(6,100);
        System.out.println(ATOMIC_INTEGER_ARRAY.get(6));
        System.out.println(ATOMIC_INTEGER_ARRAY.get(15));
        System.out.println(ATOMIC_INTEGER_ARRAY.compareAndSet(6,
                100, 200));
        System.out.println(ATOMIC_INTEGER_ARRAY.get(6));
    }
}
```

执行结果如下：

```
100
0
true
200
```

4. length()

返回当前数组的长度。

6.6 LongAdder

底层维护多个 long 型字段,并且可以动态增长以减少 CAS 争用,动态增长受 CPU 核心数量限制。在高并发场景下,此类比 AtomicLong 更可取,所占用资源相对也会更多。

6.6.1 实现方式

核心实现方式封装在 Striped64 类中,LongAdder 继承了此类。

1. 核心字段

Striped64 核心字段说明见表 6-5。

表 6-5　Striped64 核心字段

名称	类型	描述
NCPU	int	CPU 核心数量
cells	Cell[]	数组会动态扩容,最大长度受 CPU 核心数量限制。Cell 类里包装一个 long 型字段,底层基于 VarHandle 实现原子操作
base	long	基本的一个数据存储字段
cellsBusy	int	标记位,表示当前是否有执行线程正在操作 cells 数组添加数据、扩容、初始化

核心字段基于 VarHandle 实现原子操作,如图 6-10 所示。

```
//Striped64类
private static final VarHandle BASE;
private static final VarHandle CELLSBUSY;
private static final VarHandle THREAD_PROBE;
static {
    try {
        MethodHandles.Lookup l = MethodHandles.lookup();
        BASE = l.findVarHandle(Striped64.class,
                "base", long.class);
        CELLSBUSY = l.findVarHandle(Striped64.class,
                "cellsBusy", int.class);
        l = java.security.AccessController.doPrivileged(
                new java.security.PrivilegedAction<>() {
                    public MethodHandles.Lookup run() {
                        try {
                            return MethodHandles.privateLookupIn(Thread.class, MethodHandles.lookup());
                        } catch (ReflectiveOperationException e) {
                            throw new ExceptionInInitializerError(e);
                        }
                    }});
        THREAD_PROBE = l.findVarHandle(Thread.class,
                "threadLocalRandomProbe", int.class);
    } catch (ReflectiveOperationException e) {
        throw new ExceptionInInitializerError(e);
    }
}
```

图 6-10　Striped64 类源代码

2. 核心方法

longAccumulate(long x, LongBinaryOperator fn, boolean wasUncontended, int index)
如图 6-11 所示。

```java
//Striped64类
final void longAccumulate(long x, LongBinaryOperator fn,
                          boolean wasUncontended, int index) {
    if (index == 0) {
        ThreadLocalRandom.current(); //强制初始化当前执行线程对象的探测值
        index = getProbe();
        wasUncontended = true;
    }
    for (boolean collide = false;;) { //死循环,直到数据保存成功
        Cell[] cs; Cell c; int n; long v;
        if ((cs = cells) != null && (n = cs.length) > 0) {
            if ((c = cs[(n - 1) & index]) == null) {
                if (cellsBusy == 0) {       //标记位
                    Cell r = new Cell(x);
                    if (cellsBusy == 0 && casCellsBusy()) {//标记位==0并且将标记位修改为1
                        try {
                            Cell[] rs; int m, j;
                            if ((rs = cells) != null &&
                                (m = rs.length) > 0 &&
                                rs[j = (m - 1) & index] == null) {
                                rs[j] = r;//索引位存储新数据
                                break;//退出循环
                            }
                        } finally {
                            cellsBusy = 0;//标记位重置
                        }
                        continue;
                    }
                }
                collide = false;
            }
            else if (!wasUncontended)
                wasUncontended = true;
            else if (c.cas(v = c.value,
                    (fn == null) ? v + x : fn.applyAsLong(v, x)))//索引位修改数据
                break;//退出循环
            else if (n >= NCPU || cells != cs)//扩容受NCPU限制
                collide = false;
            else if (!collide)
                collide = true;
            else if (cellsBusy == 0 && casCellsBusy()) {
                try {
                    if (cells == cs)        //扩容
                        cells = Arrays.copyOf(cs, n << 1);
                } finally {
                    cellsBusy = 0;
                }
                collide = false;
                continue;
            }
            index = advanceProbe(index);//重新生成当前执行线程对象的探测值
        }
        else if (cellsBusy == 0 && cells == cs && casCellsBusy()) {
            try {                           //初始化数组
                if (cells == cs) {
                    Cell[] rs = new Cell[2];
                    rs[index & 1] = new Cell(x);
                    cells = rs;
                    break;//退出循环
                }
            } finally {
                cellsBusy = 0;
            }
        }
        else if (casBase(v = base,
                (fn == null) ? v + x : fn.applyAsLong(v, x)))//修改数据存储在base字段
            break;//退出循环
    }
}
```

图 6-11 longAccumulate 方法源代码

6.6.2 常用方法

1. add(long x)

增加给定值，接收 long 入参，作为给定值，代码如下：

```java
//第6章/six/LongAdderTest.java
public class LongAdderTest {

    private static final LongAdder LONG_ADDER = new LongAdder();

    public static void main(String[] args) {
        LONG_ADDER.add(100);
        LONG_ADDER.add(100);
        LONG_ADDER.add(100);
        System.out.println(LONG_ADDER.sum()); //300
    }
}
```

执行结果如下：

```
300
```

2. decrement()

减小给定值，等效于 add(-1)。

3. increment()

增加给定值，等效于 add(1)。

4. reset()

将所有数据重置为 0。

5. sum()

返回当前数据总和，源代码如图 6-12 所示。

6. sumThenReset()

先返回当前数据总和，再将所有数据重置为 0。

7. doubleValue()

将 sum() 结果转换为 double 值返回。

```java
public long sum() {
    Cell[] cs = cells;
    long sum = base;
    if (cs != null) {
        for (Cell c : cs)
            if (c != null)
                sum += c.value;
    }
    return sum;
}
```

图 6-12 sum 方法源代码

小结

原子操作比锁粒度更小，适用于更新某个具体的值，如 int、long、对象引用等。多个原子操作配合起来也可以达到锁的效果，但是设计难度及复杂度更高。

习题

1. 判断题

（1）原子操作比锁粒度更小。（　　）

（2）AtomicInteger 对象可以并发安全地修改底层包装的 int 值。（　　）

（3）AtomicReference<V>对象可以并发安全地修改底层包装的对象引用值。（　　）

（4）CAS 操作类似于乐观锁概念。（　　）

2. 选择题

（1）原子包是（　　）版本推出的。（单选）

  A. 1.5　　　　　　B. 1.6　　　　　　C. 1.7　　　　　　D. 1.8

（2）VarHandle 是（　　）版本推出的。（单选）

  A. 1.8　　　　　　B. 1.9　　　　　　C. 11　　　　　　D. 13

（3）以下（　　）类的底层实现基于 VarHandle。（多选）

  A. AtomicLong　　　　　　　　　　　　B. AtomicInteger

  C. AtomicReference<V>　　　　　　　　D. AtomicIntegerArray

3. 填空题

（1）根据业务要求补全代码，多线程并发操作数据，要求输出的结果为 200 000，代码如下：

```java
//第 6 章/answer/AtomicCount.java
public class AtomicCount {
    public static final AtomicInteger ATOMIC_INTEGER_ADD =
                                        new AtomicInteger();

    public static void main(String[] args) throws Exception {
        RunIncrement runIncrement = new RunIncrement();
        Thread thread = new Thread(runIncrement);
        Thread thread1 = new Thread(runIncrement);
        thread.start();
        thread1.start();
        _____;
        _____;
        System.out.println("原子:" + ATOMIC_INTEGER_ADD.get());
    }

    static final class RunIncrement implements Runnable{

        @Override
        public void run() {
            for (int i = 0; i < 100000; i++) {
                _____;
```

```
            }
        }
    }
}
```

(2) 根据业务要求补全代码，CAS 操作要求输出结果为 10，代码如下：

```
//第 6 章/answer/AtomicCAS.java
public class AtomicCAS {
    public static final AtomicInteger ATOMIC_INTEGER_ADD =
                                            new AtomicInteger(2);

    public static void main(String[] args) throws Exception {
        RunIncrement runIncrement = new RunIncrement();
        Thread thread = new Thread(runIncrement);
        thread.start();
        thread.join();
        System.out.println("原子:" + ATOMIC_INTEGER_ADD.get());
    }

    static final class RunIncrement implements Runnable{

        @Override
        public void run() {
            while (_____){

            }
        }
    }
}
```

第 7 章 阻塞队列

BlockingQueue 阻塞队列接口，常用于生产者、消费者模型中。BlockingQueue 方法有 4 种形式的处理操作方法，见表 7-1。

表 7-1　BlockingQueue 方法的 4 种形式

	引发异常	返回数据	阻塞	最大等待时间
Insert	add(e)	offer(e)	put(e)	offer(e,time,unit)
Remove	remove()	poll()	take()	poll(time,unit)
Examine	element()	peek()		

7.1　ArrayBlockingQueue

由数组支持的有界阻塞队列，此队列对元素 FIFO（先进先出）进行排序，支持可选的公平策略。新元素插入队列尾部，队列检索操作获取队列头部的元素。

7.1.1　构造器

ArrayBlockingQueue 构造器见表 7-2。

表 7-2　ArrayBlockingQueue 构造器

构造器	描述
ArrayBlockingQueue(int capacity)	构造新的对象，指定数组长度
ArrayBlockingQueue(int capacity,boolean fair)	构造新的对象，指定数组长度，指定公平策略
ArrayBlockingQueue(int capacity, boolean fair, Collection<? extends E> c)	构造新的对象，指定数组长度，指定公平策略，指定初始数据

7.1.2　常用方法

1. add(E e)

添加数据，接收泛型入参，作为新加数据。如果可以立即添加指定的数据而不会超出队列的容量，则在此队列的尾部添加指定的数据并返回 true；如果此队列已满，则引发

IllegalStateException 异常,代码如下:

```java
//第 7 章/one/ArrayBlockingQueueTest.java
public class ArrayBlockingQueueTest {

    private static final ArrayBlockingQueue<String> ARRAY_BLOCKING_QUEUE =
                                    new ArrayBlockingQueue<>(8);

    public static void main(String[] args) throws InterruptedException {
        System.out.println(ARRAY_BLOCKING_QUEUE.add("one"));
    }
}
```

执行结果如下:

```
true
```

观察源代码,会发现底层调用 offer(e)方法添加数据,失败后抛出异常,如图 7-1 所示。

```java
public boolean add(E e) {
    if (offer(e))
        return true;
    else
        throw new IllegalStateException("Queue full");
}
```

图 7-1 add 方法源代码

2. remove()

检索并删除此队列的头部数据。如果检索成功,则返回删除的数据;如果此队列为空,则引发 NoSuchElementException 异常,代码如下:

```java
//第 7 章/one/ArrayBlockingQueueTest.java
public class ArrayBlockingQueueTest {

    private static final ArrayBlockingQueue<String> ARRAY_BLOCKING_QUEUE =
                                    new ArrayBlockingQueue<>(8);

    public static void main(String[] args) throws InterruptedException {
        System.out.println(ARRAY_BLOCKING_QUEUE.add("one"));
        System.out.println(ARRAY_BLOCKING_QUEUE.remove());
    }
}
```

执行结果如下:

```
true
one
```

观察源代码,会发现最终调用 poll()方法删除数据,失败后抛出异常,如图 7-2 所示。

```
public E remove() {
    E x = poll();
    if (x != null)
        return x;
    else
        throw new NoSuchElementException();
}
```

图 7-2 remove 方法源代码

3. element()

检索但不删除此队列的头部数据。如果成功,则返回检索的数据;如果此队列为空,则引发 NoSuchElementException 异常,代码如下:

```java
//第7章/one/ArrayBlockingQueueTest.java
public class ArrayBlockingQueueTest {

    private static final ArrayBlockingQueue<String> ARRAY_BLOCKING_QUEUE =
                                        new ArrayBlockingQueue<>(8);

    public static void main(String[] args) throws InterruptedException {
        System.out.println(ARRAY_BLOCKING_QUEUE.add("one"));
        System.out.println(ARRAY_BLOCKING_QUEUE.element());
        System.out.println(ARRAY_BLOCKING_QUEUE.element());
    }
}
```

执行结果如下:

```
true
one
one
```

观察源代码,会发现底层调用 peek() 方法检索数据,失败后抛出异常,如图 7-3 所示。

```
public E element() {
    E x = peek();
    if (x != null)
        return x;
    else
        throw new NoSuchElementException();
}
```

图 7-3 element 方法源代码

4. offer(e)

添加数据,接收泛型入参,作为新加数据。如果可以立即添加指定的数据而不会超出队列的容量,则在此队列的尾部添加指定的数据并返回 true;如果此队列已满,则返回 false,代码如下:

```
//第7章/one/ArrayBlockingQueueTest.java
public class ArrayBlockingQueueTest {

    private static final ArrayBlockingQueue<String> ARRAY_BLOCKING_QUEUE =
                            new ArrayBlockingQueue<>(8);

    public static void main(String[] args) throws InterruptedException {
        System.out.println(ARRAY_BLOCKING_QUEUE.offer("one"));
        System.out.println(ARRAY_BLOCKING_QUEUE.offer("two"));
    }
}
```

执行结果如下：

```
true
true
```

观察源代码,会发现底层使用ReentrantLock锁,并配备了类似生产和消费的唤醒操作,如图7-4所示。

```
//ArrayBlockingQueue
public boolean offer(E e) {
    Objects.requireNonNull(e);
    final ReentrantLock lock = this.lock;
    lock.lock();    ← 加锁
    try {
        if (count == items.length)
            return false;    ← 容量满，返回false
        else {
            enqueue(e);
            return true;
        }
    } finally {
        lock.unlock();    ← 释放锁
    }
}

private void enqueue(E e) {
    final Object[] items = this.items;
    items[putIndex] = e;
    if (++putIndex == items.length) putIndex = 0;
    count++;    ← 数量标记
    notEmpty.signal();    ← 唤醒消费者
}
```

图7-4 offer方法源代码

5. poll()

检索并删除此队列的头部数据。如果检索成功,则返回删除的数据；如果此队列为空,则返回null,代码如下：

```java
//第 7 章/one/ArrayBlockingQueueTest.java
public class ArrayBlockingQueueTest {

    private static final ArrayBlockingQueue<String> ARRAY_BLOCKING_QUEUE =
                                    new ArrayBlockingQueue<>(8);

    public static void main(String[] args) throws InterruptedException {
        System.out.println(ARRAY_BLOCKING_QUEUE.offer("one"));
        System.out.println(ARRAY_BLOCKING_QUEUE.poll());
        System.out.println(ARRAY_BLOCKING_QUEUE.poll());
    }
}
```

执行结果如下：

```
true
one
null
```

观察源代码，会发现底层使用 ReentrantLock 锁，并配备了类似生产和消费的唤醒操作，如图 7-5 所示。

```
//ArrayBlockingQueue
public E poll() {
    final ReentrantLock lock = this.lock;
    lock.lock();        ← 加锁
    try {
        return (count == 0) ? null : dequeue();
    } finally {
        lock.unlock();  ← 释放锁
    }
}
private E dequeue() {
    final Object[] items = this.items;
    @SuppressWarnings("unchecked")
    E e = (E) items[takeIndex];
    items[takeIndex] = null;
    if (++takeIndex == items.length) takeIndex = 0;
    count--;            ← 数量标记
    if (itrs != null)
        itrs.elementDequeued();
    notFull.signal();   ← 唤醒生产者
    return e;
}
```

图 7-5　poll 方法源代码

6. peek()

检索但不删除此队列的头部数据。如果成功，则返回检索的数据；如果此队列为空，则

返回 null,代码如下:

```java
//第 7 章/one/ArrayBlockingQueueTest.java
public class ArrayBlockingQueueTest {

    private static final ArrayBlockingQueue<String> ARRAY_BLOCKING_QUEUE =
                                        new ArrayBlockingQueue<>(8);

    public static void main(String[] args) throws InterruptedException {
        System.out.println(ARRAY_BLOCKING_QUEUE.offer("one"));
        System.out.println(ARRAY_BLOCKING_QUEUE.peek());
        System.out.println(ARRAY_BLOCKING_QUEUE.peek());
    }
}
```

执行结果如下:

```
true
one
one
```

观察源代码,会发现底层使用 ReentrantLock 锁,返回 takeIndex 索引处数据,如图 7-6 所示。

```java
//ArrayBlockingQueue
public E peek() {
    final ReentrantLock lock = this.lock;
    lock.lock();
    try {
        return itemAt(takeIndex);
    } finally {
        lock.unlock();
    }
}
final E itemAt(int i) {
    return (E) items[i];
}
```

图 7-6　peek 方法源代码

7. put(e)

添加数据,接收泛型入参,作为新加数据。如果可以立即添加指定的数据而不会超出队列的容量,则在此队列的尾部添加指定的数据;如果此队列已满,则阻塞等待,代码如下:

```java
//第 7 章/one/ArrayBlockingQueueTest.java
public class ArrayBlockingQueueTest {

    private static final ArrayBlockingQueue<String> ARRAY_BLOCKING_QUEUE =
                                        new ArrayBlockingQueue<>(8);
```

```java
public static void main(String[] args) throws InterruptedException {
    ARRAY_BLOCKING_QUEUE.put("one");
}
}
```

观察源代码,会发现底层使用 ReentrantLock 锁,并配备了类似生产和消费的唤醒操作。当没有可用空间时 await()阻塞等待,如图 7-7 所示。

```java
public void put(E e) throws InterruptedException {
    Objects.requireNonNull(e);
    final ReentrantLock lock = this.lock;
    lock.lockInterruptibly();    ← 加锁
    try {
        while (count == items.length)
            notFull.await();    ← 阻塞等待
        enqueue(e);
    } finally {
        lock.unlock();    ← 释放锁
    }
}
```

图 7-7 put 方法源代码

8. take()

检索并删除此队列的头部数据。如果检索成功,则返回删除的数据;如果此队列为空,则阻塞等待,代码如下:

```java
//第7章/one/ArrayBlockingQueueTest.java
public class ArrayBlockingQueueTest {

    private static final ArrayBlockingQueue<String> ARRAY_BLOCKING_QUEUE =
                                    new ArrayBlockingQueue<>(8);

    public static void main(String[] args) throws InterruptedException {
        ARRAY_BLOCKING_QUEUE.put("one");
        System.out.println(ARRAY_BLOCKING_QUEUE.take());
    }
}
```

执行结果如下:

```
one
```

观察源代码,会发现底层使用 ReentrantLock 锁,并配备了类似生产和消费的唤醒操作。当数据为空时 await()阻塞等待,如图 7-8 所示。

9. offer(E e, long timeout, TimeUnit unit)

添加数据,接收泛型入参,作为新加数据;接收 long 入参,作为最大等待时间;接收 TimeUnit 入参,作为时间单位。如果可以立即添加指定的数据而不会超出队列的容量,则

```java
//ArrayBlockingQueue
public E take() throws InterruptedException {
    final ReentrantLock lock = this.lock;
    lock.lockInterruptibly();
    try {
        while (count == 0)   ← 数据为空时
            notEmpty.await();
        return dequeue();
    } finally {
        lock.unlock();
    }
}

private E dequeue() {
    final Object[] items = this.items;
    @SuppressWarnings("unchecked")
    E e = (E) items[takeIndex];
    items[takeIndex] = null;
    if (++takeIndex == items.length) takeIndex = 0;
    count--;
    if (itrs != null)
        itrs.elementDequeued();
    notFull.signal();   ← 唤醒操作
    return e;
}
```

图 7-8　take 方法源代码

在此队列的尾部添加指定的数据并返回 true；如果此队列已满，则给定最大等待时间段；如果在此时间段内有空位，则添加成功并返回 true，否则返回 false，代码如下：

```java
//第 7 章/one/ArrayBlockingQueueTest.java
public class ArrayBlockingQueueTest {

    private static final ArrayBlockingQueue<String> ARRAY_BLOCKING_QUEUE =
                                    new ArrayBlockingQueue<>(8);

    public static void main(String[] args) throws InterruptedException {
        ARRAY_BLOCKING_QUEUE.offer("one",2, TimeUnit.SECONDS);
        System.out.println(ARRAY_BLOCKING_QUEUE.take());
    }
}
```

执行结果如下：

```
one
```

观察源代码，会发现底层使用 ReentrantLock 锁，并配备了类似生产和消费的唤醒操作。当没有可用空间时 awaitNanos(long nanosTimeout) 阻塞等待最大给定时间段，如图 7-9 所示。

```
public boolean offer(E e, long timeout, TimeUnit unit)
        throws InterruptedException {

    Objects.requireNonNull(e);
    long nanos = unit.toNanos(timeout);    ← 时间格式转换
    final ReentrantLock lock = this.lock;
    lock.lockInterruptibly();
    try {
        while (count == items.length) {
            if (nanos <= 0L)
                return false;
            nanos = notFull.awaitNanos(nanos);
        }                                   ← 给定最大等待时间段
        enqueue(e);
        return true;
    } finally {
        lock.unlock();
    }
}
```

图 7-9　offer 方法源代码

10. poll(long timeout, TimeUnit unit)

检索并删除此队列的头部数据。接收 long 入参,作为最大等待时间；接收 TimeUnit 入参,作为时间单位。如果检索成功,则返回删除的数据；如果此队列为空,则给定最大等待时间段；在此时间段内如果有数据,则继续检索操作,否则返回 null,代码如下：

```java
//第 7 章/one/ArrayBlockingQueueTest.java
public class ArrayBlockingQueueTest {

    private static final ArrayBlockingQueue< String > ARRAY_BLOCKING_QUEUE =
                                    new ArrayBlockingQueue<>(8);

    public static void main(String[] args) throws InterruptedException {
        ARRAY_BLOCKING_QUEUE.offer("one",2, TimeUnit.SECONDS);
        System.out.println(ARRAY_BLOCKING_QUEUE.poll(2,
                                    TimeUnit.SECONDS));
    }
}
```

执行结果如下：

```
one
```

观察源代码,会发现底层使用 ReentrantLock 锁,并配备了类似生产和消费的唤醒操作。当数据为空时 awaitNanos(long nanosTimeout)阻塞等待最大给定时间段,如图 7-10 所示。

11. contains(Object o)

如果此队列包含指定的元素,则返回 true,否则返回 false。接收 Object 入参,作为比较

```
public E poll(long timeout, TimeUnit unit) throws InterruptedException {
    long nanos = unit.toNanos(timeout);  ← 时间格式转换
    final ReentrantLock lock = this.lock;
    lock.lockInterruptibly();
    try {
        while (count == 0) {  ← 数据为空时
            if (nanos <= 0L)
                return null;
            nanos = notEmpty.awaitNanos(nanos);  ← 给定最大等待时间段
        }
        return dequeue();
    } finally {
        lock.unlock();
    }
}
```

图 7-10　poll 方法源代码

的元素。

12．drainTo(Collection<? super E> c)

删除此队列中所有可用元素，并将它们添加到给定集合中，然后返回添加的数量。接收 Collection 入参，作为给定集合，代码如下：

```java
//第 7 章/one/ArrayBlockingQueueTest.java
public class ArrayBlockingQueueTest {

    private static final ArrayBlockingQueue<String> ARRAY_BLOCKING_QUEUE =
                                    new ArrayBlockingQueue<>(8);

    public static void main(String[] args) throws InterruptedException {
        ARRAY_BLOCKING_QUEUE.put("one");
        ARRAY_BLOCKING_QUEUE.put("two");
        System.out.println(ARRAY_BLOCKING_QUEUE.contains("one"));
        ArrayList<String> arrayList = new ArrayList<>();
        System.out.println(ARRAY_BLOCKING_QUEUE.drainTo(arrayList));
        System.out.println(arrayList);
        System.out.println(ARRAY_BLOCKING_QUEUE); //此对象的 toString()方法
    }
}
```

执行结果如下：

```
true
2
[one, two]
[]
```

13．forEach(Consumer<? super E> action)

对可迭代的每个元素执行给定的操作，直到处理完所有元素或操作引发异常。接收

Consumer 入参,作为给定的操作对象,代码如下:

```
//第 7 章/one/ArrayBlockingQueueTest.java
public class ArrayBlockingQueueTest {

    private static final ArrayBlockingQueue<String> ARRAY_BLOCKING_QUEUE =
                                    new ArrayBlockingQueue<>(8);

    public static void main(String[] args) throws InterruptedException {
        ARRAY_BLOCKING_QUEUE.put("one");
        ARRAY_BLOCKING_QUEUE.put("two");
        ARRAY_BLOCKING_QUEUE.forEach(System.out::println);
    }
}
```

执行结果如下:

```
one
two
```

14. remainingCapacity()

返回此队列在理想情况下还可以接收的元素数量,代码如下:

```
//第 7 章/one/ArrayBlockingQueueTest.java
public class ArrayBlockingQueueTest {

    private static final ArrayBlockingQueue<String> ARRAY_BLOCKING_QUEUE =
                                    new ArrayBlockingQueue<>(8);

    public static void main(String[] args) throws InterruptedException {
        ARRAY_BLOCKING_QUEUE.put("one");
        ARRAY_BLOCKING_QUEUE.put("two");
        System.out.println(ARRAY_BLOCKING_QUEUE.remainingCapacity());
    }
}
```

执行结果如下:

```
6
```

15. remove(Object o)

删除指定元素数据,如果匹配元素成功,则删除,返回值为 true,否则返回值为 false。接收 Object 入参,作为指定元素对象。

16. removeAll(Collection<?> c)

删除此集合中的所有元素,如果匹配任意元素成功,则删除,返回值为 true,否则返回值为 false。接收 Collection 入参,作为指定集合,代码如下:

```java
public class ArrayBlockingQueueTest {

    private static final ArrayBlockingQueue<String> ARRAY_BLOCKING_QUEUE =
                                    new ArrayBlockingQueue<>(8);

    public static void main(String[] args) throws InterruptedException {
        ARRAY_BLOCKING_QUEUE.put("one");
        ARRAY_BLOCKING_QUEUE.put("two");
        List<String> one = List.of("one");
        System.out.println(ARRAY_BLOCKING_QUEUE.removeAll(one));
        System.out.println(ARRAY_BLOCKING_QUEUE); //此对象的 toString()方法
    }
}
```

执行结果如下：

```
true
[two]
```

17. removeIf(Predicate<? super E> filter)

对可迭代的每个元素执行给定的操作，如果返回值为 true，则表示删除此数据；如果返回值为 false，则表示不删除此数据。接收 Predicate 入参，作为给定的操作对象，代码如下：

```java
//第 7 章/one/ArrayBlockingQueueTest.java
public class ArrayBlockingQueueTest {

    private static final ArrayBlockingQueue<String> ARRAY_BLOCKING_QUEUE =
                                    new ArrayBlockingQueue<>(8);

    public static void main(String[] args) throws InterruptedException {
        ARRAY_BLOCKING_QUEUE.put("one");
        ARRAY_BLOCKING_QUEUE.put("two");
        System.out.println(ARRAY_BLOCKING_QUEUE.removeIf((e) -> {
            System.out.println("removeIf:" + e);
            return "one".equals(e);                    //删除 one 数据
        }));
        System.out.println(ARRAY_BLOCKING_QUEUE);      //此对象的 toString()方法
    }
}
```

执行结果如下：

```
removeIf:one
removeIf:two
true
[two]
```

18. size()

返回此队列中的已添加元素的数量，代码如下：

```java
//第 7 章/one/ArrayBlockingQueueTest.java
public class ArrayBlockingQueueTest {

    private static final ArrayBlockingQueue<String> ARRAY_BLOCKING_QUEUE =
                                    new ArrayBlockingQueue<>(8);

    public static void main(String[] args) throws InterruptedException {
        ARRAY_BLOCKING_QUEUE.put("one");
        ARRAY_BLOCKING_QUEUE.put("two");
        System.out.println(ARRAY_BLOCKING_QUEUE.size());
    }
}
```

执行结果如下:

```
2
```

7.2 LinkedBlockingQueue

由单向链表支持的有界阻塞队列,此队列对元素 FIFO(先进先出)进行排序。新元素插入队列尾部,队列检索操作获取队列头部的元素。核心字段如图 7-11 所示。

```java
/** 容量限制,如果没有,则为整数.MAX_VALUE */
private final int capacity;
/** 当前元素数量 */
private final AtomicInteger count = new AtomicInteger();

 /** 链表头 */
transient Node<E> head;

/** 链表尾 */
private transient Node<E> last;

/** 消费锁 */
private final ReentrantLock takeLock = new ReentrantLock();

@SuppressWarnings("serial")
private final Condition notEmpty = takeLock.newCondition();

/** 生产锁 */
private final ReentrantLock putLock = new ReentrantLock();

@SuppressWarnings("serial")
private final Condition notFull = putLock.newCondition();
```

图 7-11 LinkedBlockingQueue 核心字段

7.2.1 构造器

LinkedBlockingQueue 构造器见表 7-3。

表 7-3 LinkedBlockingQueue 构造器

构造器	描述
LinkedBlockingQueue()	构造新的对象,默认为无参构造器
LinkedBlockingQueue(int capacity)	构造新的对象,指定链表长度
LinkedBlockingQueue(Collection<? extends E> c)	构造新的对象,指定初始数据

7.2.2 常用方法

1. add(E e)

添加数据,接收泛型入参,作为新加数据。如果可以立即添加指定的数据而不会超出队列的容量,则在此队列的尾部添加指定的数据并返回 true;如果此队列已满,则引发 IllegalStateException 异常,代码如下:

```java
//第 7 章/two/LinkedBlockingQueueTest.java
public class LinkedBlockingQueueTest {

    private static final LinkedBlockingQueue<String> BLOCKING_QUEUE =
                                    new LinkedBlockingQueue<>(8);

    public static void main(String[] args) throws InterruptedException {
        System.out.println(BLOCKING_QUEUE.add("one"));
    }
}
```

执行结果如下:

```
true
```

观察源代码,会发现底层调用 offer(e)方法添加数据,失败后抛出异常,如图 7-12 所示。

```java
public boolean add(E e) {
    if (offer(e))
        return true;
    else
        throw new IllegalStateException("Queue full");
}
```

图 7-12 add 方法源代码

2. remove()

检索并删除此队列的头部数据。如果检索成功,则返回删除的数据,如果此队列为空,则引发 NoSuchElementException 异常,代码如下:

```java
//第 7 章/two/LinkedBlockingQueueTest.java
public class LinkedBlockingQueueTest {

    private static final LinkedBlockingQueue<String> BLOCKING_QUEUE =
                                            new LinkedBlockingQueue<>(8);

    public static void main(String[] args) throws InterruptedException {
        System.out.println(BLOCKING_QUEUE.add("one"));
        System.out.println(BLOCKING_QUEUE.remove());
    }
}
```

执行结果如下:

```
true
one
```

观察源代码,会发现底层调用 poll() 方法删除数据,失败后抛出异常,如图 7-13 所示。

```java
public E remove() {
    E x = poll();
    if (x != null)
        return x;
    else
        throw new NoSuchElementException();
}
```

图 7-13　remove 方法源代码

3. element()

检索但不删除此队列的头部数据。如果成功,则返回检索的数据;如果此队列为空,则引发 NoSuchElementException 异常,代码如下:

```java
//第 7 章/two/LinkedBlockingQueueTest.java
public class LinkedBlockingQueueTest {

    private static final LinkedBlockingQueue<String> BLOCKING_QUEUE =
                                            new LinkedBlockingQueue<>(8);

    public static void main(String[] args) throws InterruptedException {
        System.out.println(BLOCKING_QUEUE.add("one"));
        System.out.println(BLOCKING_QUEUE.element());
        System.out.println(BLOCKING_QUEUE.element());
    }
}
```

执行结果如下:

```
true
one
one
```

观察源代码，会发现底层调用 peek() 方法检索数据，失败后抛出异常，如图 7-14 所示。

```
public E element() {
    E x = peek();
    if (x != null)
        return x;
    else
        throw new NoSuchElementException();
}
```

图 7-14 element 方法源代码

4. offer(E e)

添加数据，接收泛型入参，作为新加数据。如果可以立即添加指定的数据而不会超出队列的容量，则在此队列的尾部添加指定的数据并返回 true；如果此队列已满，则返回 false，代码如下：

```java
//第 7 章/two/LinkedBlockingQueueTest.java
public class LinkedBlockingQueueTest {

    private static final LinkedBlockingQueue<String> BLOCKING_QUEUE =
                                    new LinkedBlockingQueue<>(8);

    public static void main(String[] args) throws InterruptedException {
        System.out.println(BLOCKING_QUEUE.offer("one"));
    }
}
```

执行结果如下：

```
true
```

观察源代码，会发现底层使用 ReentrantLock 锁，并配备了类似生产和消费的唤醒操作，如图 7-15 所示。

5. poll()

检索并删除此队列的头部数据。如果检索成功，则返回删除的数据；如果此队列为空，则返回 null，代码如下：

```java
//第 7 章/two/LinkedBlockingQueueTest.java
public class LinkedBlockingQueueTest {

    private static final LinkedBlockingQueue<String> BLOCKING_QUEUE =
                                    new LinkedBlockingQueue<>(8);

    public static void main(String[] args) throws InterruptedException {
        BLOCKING_QUEUE.put("one");
        System.out.println(BLOCKING_QUEUE.poll());
        System.out.println(BLOCKING_QUEUE.poll());
    }
}
```

```java
public boolean offer(@NotNull E e) {
    if (e == null) throw new NullPointerException();
    final AtomicInteger count = this.count;
    if (count.get() == capacity)
        return false;  // 容量满
    final int c;
    final Node<E> node = new Node<E>(e);
    final ReentrantLock putLock = this.putLock;
    putLock.lock();  // 加锁
    try {
        if (count.get() == capacity)
            return false;  // 容量满
        enqueue(node);  // 添加数据
        c = count.getAndIncrement();  // 原子更新数量
        if (c + 1 < capacity)
            notFull.signal();  // 唤醒生产者
    } finally {
        putLock.unlock();  // 释放锁
    }
    if (c == 0)
        signalNotEmpty();  // 唤醒消费者
    return true;
}
```

图 7-15 offer 方法源代码

执行结果如下:

```
one
null
```

观察源代码,会发现底层使用 ReentrantLock 锁,并配备了类似生产和消费的唤醒操作,如图 7-16 所示。

6. peek()

检索但不删除此队列的头部数据。如果成功,则返回检索的数据;如果此队列为空,则返回 null,代码如下:

```java
//第 7 章/two/LinkedBlockingQueueTest.java
public class LinkedBlockingQueueTest {

    private static final LinkedBlockingQueue<String> BLOCKING_QUEUE =
            new LinkedBlockingQueue<>(8);

    public static void main(String[] args) throws InterruptedException {
        BLOCKING_QUEUE.put("one");
        System.out.println(BLOCKING_QUEUE.peek());
        System.out.println(BLOCKING_QUEUE.peek());
    }
}
```

```java
public E poll() {
    final AtomicInteger count = this.count;
    if (count.get() == 0)
        return null;
    final E x;
    final int c;
    final ReentrantLock takeLock = this.takeLock;
    takeLock.lock();   // 加锁
    try {
        if (count.get() == 0)   // 数据为空时
            return null;
        x = dequeue();   // 数据出队列
        c = count.getAndDecrement();
        if (c > 1)
            notEmpty.signal();   // 唤醒消费者
    } finally {
        takeLock.unlock();   // 释放锁
    }
    if (c == capacity)
        signalNotFull();   // 唤醒生产者
    return x;
}
```

图 7-16 poll 方法源代码

执行结果如下：

```
one
one
```

观察源代码，会发现底层使用 ReentrantLock 锁，返回链表头部索引处数据，如图 7-17 所示。

```java
public E peek() {
    final AtomicInteger count = this.count;
    if (count.get() == 0)
        return null;
    final ReentrantLock takeLock = this.takeLock;
    takeLock.lock();
    try {
        return (count.get() > 0) ? head.next.item : null;   // 头部数据
    } finally {
        takeLock.unlock();
    }
}
```

图 7-17 peek 方法源代码

7．put(E e)

添加数据，接收泛型入参，作为新加数据。如果可以立即添加指定的数据而不会超出队列的容量，则在此队列的尾部添加指定的数据；如果此队列已满，则阻塞等待，代码如下：

```java
//第 7 章/two/LinkedBlockingQueueTest.java
public class LinkedBlockingQueueTest {

    private static final LinkedBlockingQueue<String> BLOCKING_QUEUE =
                                    new LinkedBlockingQueue<>(8);

    public static void main(String[] args) throws InterruptedException {
        BLOCKING_QUEUE.put("one");
        BLOCKING_QUEUE.put("two");
        System.out.println(BLOCKING_QUEUE.peek());
    }
}
```

执行结果如下:

```
one
```

观察源代码,会发现底层使用 ReentrantLock 锁,并配备了类似生产和消费的唤醒操作。当没有可用空间时 await()阻塞等待,如图 7-18 所示。

```
public void put(E e) throws InterruptedException {
    if (e == null) throw new NullPointerException();
    final int c;
    final Node<E> node = new Node<E>(e);    ← 数据节点
    final ReentrantLock putLock = this.putLock;
    final AtomicInteger count = this.count;
    putLock.lockInterruptibly();    ← 加锁
    try {
        while (count.get() == capacity) {
            notFull.await();    ← 数据已满
        }
        enqueue(node);    ← 添加数据
        c = count.getAndIncrement();    ← 原子操作
        if (c + 1 < capacity)
            notFull.signal();    ← 唤醒生产者
    } finally {
        putLock.unlock();    ← 释放锁
    }
    if (c == 0)
        signalNotEmpty();    ← 唤醒消费者
}
```

图 7-18   put 方法源代码

8. take()

检索并删除此队列的头部数据。如果检索成功,则返回删除的数据;如果此队列为空,则阻塞等待,代码如下:

```java
//第7章/two/LinkedBlockingQueueTest.java
public class LinkedBlockingQueueTest {

    private static final LinkedBlockingQueue<String> BLOCKING_QUEUE =
                                    new LinkedBlockingQueue<>(8);

    public static void main(String[] args) throws InterruptedException {
        BLOCKING_QUEUE.put("one");
        BLOCKING_QUEUE.put("two");
        System.out.println(BLOCKING_QUEUE.take());
        System.out.println(BLOCKING_QUEUE.take());
    }
}
```

执行结果如下：

```
one
two
```

观察源代码，会发现底层使用 ReentrantLock 锁，并配备了类似生产和消费的唤醒操作。当数据为空时 await() 阻塞等待，如图 7-19 所示。

```
public E take() throws InterruptedException {
    final E x;
    final int c;
    final AtomicInteger count = this.count;
    final ReentrantLock takeLock = this.takeLock;
    takeLock.lockInterruptibly();        ← 加锁
    try {
        while (count.get() == 0) {
            notEmpty.await();             ← 数据为空时
        }
        x = dequeue();                    ← 数据出队列
        c = count.getAndDecrement();      ← 原子更新数量
        if (c > 1)
            notEmpty.signal();            ← 唤醒消费者
    } finally {
        takeLock.unlock();                ← 释放锁
    }
    if (c == capacity)
        signalNotFull();                  ← 唤醒生产者
    return x;
}
```

图 7-19 take 方法源代码

9. offer(E e, long timeout, TimeUnit unit)

添加数据，接收泛型入参，作为新加数据；接收 long 入参，作为最大等待时间；接收 TimeUnit 入参，作为时间单位。如果可以立即添加指定的数据而不会超出队列的容量，则在此队列的尾部添加指定的数据并返回 true；如果此队列已满，则最大等待给定时间段；在

此时间段内如果有空位,则添加成功并返回 true,否则返回 false,代码如下:

```java
//第 7 章/two/LinkedBlockingQueueTest.java
public class LinkedBlockingQueueTest {

    private static final LinkedBlockingQueue<String> BLOCKING_QUEUE =
                                        new LinkedBlockingQueue<>(8);

    public static void main(String[] args) throws InterruptedException {
        System.out.println(BLOCKING_QUEUE.offer("one", 2,
                TimeUnit.SECONDS));
        System.out.println(BLOCKING_QUEUE.offer("two", 2,
                TimeUnit.SECONDS));

    }
}
```

执行结果如下:

```
true
true
```

观察源代码,会发现底层使用 ReentrantLock 锁,并配备了类似生产和消费的唤醒操作。当没有可用空间时 awaitNanos(long nanosTimeout)阻塞等待最大给定时间段,如图 7-20 所示。

```java
public boolean offer(E e, long timeout, TimeUnit unit)
        throws InterruptedException {

    if (e == null) throw new NullPointerException();
    long nanos = unit.toNanos(timeout);      ← 时间格式转换
    final int c;
    final ReentrantLock putLock = this.putLock;
    final AtomicInteger count = this.count;
    putLock.lockInterruptibly();              ← 加锁
    try {
        while (count.get() == capacity) {
            if (nanos <= 0L)
                return false;
            nanos = notFull.awaitNanos(nanos);  ← 最大等待时间
        }
        enqueue(new Node<E>(e));              ← 添加数据
        c = count.getAndIncrement();          ← 原子更新数量
        if (c + 1 < capacity)
            notFull.signal();                 ← 唤醒生产者
    } finally {
        putLock.unlock();                     ← 释放锁
    }
    if (c == 0)
        signalNotEmpty();                     ← 唤醒消费者
    return true;
}
```

图 7-20 offer 方法源代码

10. poll(long timeout,TimeUnit unit)

检索并删除此队列的头部数据。接收 long 入参，作为最大等待时间；接收 TimeUnit 入参，作为时间单位。如果检索成功，则返回删除的数据；如果此队列为空，则给定最大等待时间段，在此时间段内如果有数据，则继续检索操作，否则返回 null，代码如下：

```java
//第 7 章/two/LinkedBlockingQueueTest.java
public class LinkedBlockingQueueTest {

    private static final LinkedBlockingQueue<String> BLOCKING_QUEUE =
                            new LinkedBlockingQueue<>(8);

    public static void main(String[] args) throws InterruptedException {
        System.out.println(BLOCKING_QUEUE.offer("one", 2,
                        TimeUnit.SECONDS));
        System.out.println(BLOCKING_QUEUE.poll(2, TimeUnit.SECONDS));

    }
}
```

执行结果如下：

```
true
one
```

观察源代码，会发现底层使用 ReentrantLock 锁，并配备了类似生产和消费的唤醒操作。当数据为空时 awaitNanos(long nanosTimeout)阻塞等待最大给定时间段，如图 7-21 所示。

```java
public E poll(long timeout, TimeUnit unit) throws InterruptedException {
    final E x;
    final int c;
    long nanos = unit.toNanos(timeout);      // 时间格式转换
    final AtomicInteger count = this.count;
    final ReentrantLock takeLock = this.takeLock;
    takeLock.lockInterruptibly();             // 加锁
    try {
        while (count.get() == 0) {
            if (nanos <= 0L)
                return null;
            nanos = notEmpty.awaitNanos(nanos);   // 最大等待时间
        }
        x = dequeue();                        // 数据出队列
        c = count.getAndDecrement();          // 原子更新数量
        if (c > 1)
            notEmpty.signal();                // 唤醒消费者
    } finally {
        takeLock.unlock();                    // 释放锁
    }
    if (c == capacity)
        signalNotFull();                      // 唤醒生产者
    return x;
}
```

图 7-21　poll 方法源代码

11. contains(Object o)

如果此队列包含指定的元素,则返回 true,否则返回 false。接收 Object 入参,作为比较的元素。

12. drainTo(Collection<? super E> c)

删除此队列中所有可用元素,并将它们添加到给定集合中,然后返回添加的数量。接收 Collection 入参,作为给定集合,代码如下:

```java
//第 7 章/two/LinkedBlockingQueueTest.java
public class LinkedBlockingQueueTest {

    private static final LinkedBlockingQueue<String> BLOCKING_QUEUE =
                            new LinkedBlockingQueue<>(8);

    public static void main(String[] args) throws InterruptedException {
        BLOCKING_QUEUE.put("one");
        BLOCKING_QUEUE.put("two");
        BLOCKING_QUEUE.put("three");
        List<String> stringList = new ArrayList<>();
        System.out.println(BLOCKING_QUEUE.drainTo(stringList));
        System.out.println(BLOCKING_QUEUE);
    }
}
```

执行结果如下:

```
3
[]
```

13. forEach(Consumer<? super E> action)

对可迭代的每个元素执行给定的操作,直到处理完所有元素或操作引发异常。接收 Consumer 入参,作为给定的操作对象,代码如下:

```java
//第 7 章/two/LinkedBlockingQueueTest.java
public class LinkedBlockingQueueTest {

    private static final LinkedBlockingQueue<String> BLOCKING_QUEUE =
                            new LinkedBlockingQueue<>(8);

    public static void main(String[] args) throws InterruptedException {
        BLOCKING_QUEUE.put("one");
        BLOCKING_QUEUE.put("two");
        BLOCKING_QUEUE.put("three");
        BLOCKING_QUEUE.forEach(System.out::println);
    }
}
```

执行结果如下:

```
one
two
three
```

14. remainingCapacity()

返回此队列在理想情况下还可以接收的元素数量,代码如下:

```java
//第 7 章/two/LinkedBlockingQueueTest.java
public class LinkedBlockingQueueTest {

    private static final LinkedBlockingQueue<String> BLOCKING_QUEUE =
                                    new LinkedBlockingQueue<>(8);

    public static void main(String[] args) throws InterruptedException {
        BLOCKING_QUEUE.put("one");
        BLOCKING_QUEUE.put("two");
        BLOCKING_QUEUE.put("three");
        System.out.println(BLOCKING_QUEUE.remainingCapacity());
    }
}
```

执行结果如下:

```
5
```

15. remove(Object o)

删除指定元素数据,如果匹配元素成功删除,则返回 true,否则返回 false。接收 Object 入参,作为指定元素对象。

16. removeAll(Collection<?> c)

删除此集合中的所有元素,如果匹配任意元素成功删除,则返回 true,否则返回 false。接收 Collection 入参,作为指定集合,代码如下:

```java
//第 7 章/two/LinkedBlockingQueueTest.java
public class LinkedBlockingQueueTest {

    private static final LinkedBlockingQueue<String> BLOCKING_QUEUE =
                                    new LinkedBlockingQueue<>(8);

    public static void main(String[] args) throws InterruptedException {
        BLOCKING_QUEUE.put("one");
        BLOCKING_QUEUE.put("two");
        BLOCKING_QUEUE.put("three");
        System.out.println(BLOCKING_QUEUE.removeAll(List.of("one")));
        System.out.println(BLOCKING_QUEUE); //此对象的 toString()方法
    }
}
```

执行结果如下:

```
true
[two, three]
```

17. removeIf(Predicate<? super E> filter)

对可迭代的每个元素执行给定的操作,如果返回值为 true,则表示删除此数据;如果返回值为 false,则表示不删除此数据。接收 Predicate 入参,作为给定的操作对象,代码如下:

```java
//第7章/two/LinkedBlockingQueueTest.java
public class LinkedBlockingQueueTest {

    private static final LinkedBlockingQueue<String> BLOCKING_QUEUE =
                                    new LinkedBlockingQueue<>(8);

    public static void main(String[] args) throws InterruptedException {
        BLOCKING_QUEUE.put("one");
        BLOCKING_QUEUE.put("two");
        BLOCKING_QUEUE.put("three");
        System.out.println(BLOCKING_QUEUE.removeIf("one"::equals));
        System.out.println(BLOCKING_QUEUE); //此对象的toString()方法
    }
}
```

执行结果如下:

```
true
[two, three]
```

18. size()

返回此队列中的已添加元素的数量,代码如下:

```java
//第7章/two/LinkedBlockingQueueTest.java
public class LinkedBlockingQueueTest {

    private static final LinkedBlockingQueue<String> BLOCKING_QUEUE =
                                    new LinkedBlockingQueue<>(8);

    public static void main(String[] args) throws InterruptedException {
        BLOCKING_QUEUE.put("one");
        BLOCKING_QUEUE.put("two");
        BLOCKING_QUEUE.put("three");
        System.out.println(BLOCKING_QUEUE.size());
    }
}
```

执行结果如下:

```
3
```

19. toArray()

按正确的顺序返回包含此队列中所有元素的数组，代码如下：

```java
//第7章/two/LinkedBlockingQueueTest.java
public class LinkedBlockingQueueTest {

    private static final LinkedBlockingQueue<String> BLOCKING_QUEUE =
                                    new LinkedBlockingQueue<>(8);

    public static void main(String[] args) throws InterruptedException {
        BLOCKING_QUEUE.put("one");
        BLOCKING_QUEUE.put("two");
        BLOCKING_QUEUE.put("three");
        System.out.println(Arrays.toString(BLOCKING_QUEUE.toArray()));
        System.out.println(BLOCKING_QUEUE); //此对象的 toString()方法
    }
}
```

执行结果如下：

```
[one, two, three]
[one, two, three]
```

7.3 LinkedTransferQueue

由单向链表支持的无界阻塞队列，此队列对元素 FIFO（先进先出）进行排序。新元素插入队列尾部，队列检索操作获取队列头部的元素。底层基于 VarHandle 实现原子操作，如图 7-22 所示。

```
//LinkedTransferQueue
private static final VarHandle HEAD;      ← 队列头
private static final VarHandle TAIL;      ← 队列尾
static final VarHandle ITEM;              ← 节点数据
static final VarHandle NEXT;              ← 下一个节点
static final VarHandle WAITER;            ← 阻塞等待的线程
static {
    try {
        MethodHandles.Lookup l = MethodHandles.lookup();
        HEAD = l.findVarHandle(LinkedTransferQueue.class, "head",
                        Node.class);
        TAIL = l.findVarHandle(LinkedTransferQueue.class, "tail",
                        Node.class);
        ITEM = l.findVarHandle(Node.class, "item", Object.class);
        NEXT = l.findVarHandle(Node.class, "next", Node.class);
        WAITER = l.findVarHandle(Node.class, "waiter", Thread.class);
    } catch (ReflectiveOperationException e) {
        throw new ExceptionInInitializerError(e);
    }

    Class<?> ensureLoaded = LockSupport.class;
}
```

图 7-22 LinkedTransferQueue 类源代码

Node 节点核心字段,如图 7-23 所示。

```
static final class Node implements ForkJoinPool.ManagedBlocker {
    final boolean isData;       // 是否生产数据节点
    volatile Object item;       // 数据
    volatile Node next;         // 下一个节点
    volatile Thread waiter;     // 阻塞等待的线程
```

图 7-23　Node 静态内部类核心字段

7.3.1　构造器

LinkedTransferQueue 构造器见表 7-4。

表 7-4　LinkedTransferQueue 构造器

构 造 器	描 述
LinkedTransferQueue()	构造新的对象,默认为无参构造器
LinkedTransferQueue(Collection<? extends E> c)	构造新的对象,指定初始数据

7.3.2　常用方法

底层核心方法 xfer(E e,boolean haveData,int how,long nanos),如图 7-24 所示。

```
private E xfer(E e, boolean haveData, int how, long nanos) {
    if (haveData && (e == null))                      ← 方法类型
        throw new NullPointerException();
    restart: for (Node s = null, t = null, h = null;;) {
        for (Node p = (t != (t = tail) && t.isData == haveData) ? t
                : (h = head);; ) {
            final Node q; final Object item;
            if (p.isData != haveData
                && haveData == ((item = p.item) == null)) {
                if (h == null) h = head;
                if (p.tryMatch(item, e)) {            ← 数据匹配
                    if (h != p) skipDeadNodesNearHead(h, p);
                    return (E) item;                  ← 更新head信息
                }
            }
            if ((q = p.next) == null) {
                if (how == NOW) return e;             ← 创建节点
                if (s == null) s = new Node(e);
                if (!p.casNext(null, s)) continue;    ← 节点链接
                if (p != t) casTail(t, s);
                if (how == ASYNC) return e;           ← 更新tail信息
                return awaitMatch(s, p, e, (how == TIMED), nanos);
            }                                         ← 阻塞等待
            if (p == (p = q)) continue restart;
        }
    }
}
```

图 7-24　xfer 方法源代码

方法类型有 4 种特性,如图 7-25 所示。

```
private static final int NOW   = 0; // poll, tryTransfer 立即匹配模式
private static final int ASYNC = 1; // offer, put, add 匹配消费者,当没有消费者时数据入队列
private static final int SYNC  = 2; // transfer, take 阻塞匹配模式
private static final int TIMED = 3; // poll, tryTransfer 立即匹配模式,带最大等待时间
```

图 7-25　方法类型说明

1. add(E e)

添加数据,接收泛型入参,作为新加数据。由于此类是无界队列,所以此方法的正常返回值为 true,代码如下：

```java
//第7章/three/LinkedTransferQueueTest.java
public class LinkedTransferQueueTest {

    private static final LinkedTransferQueue<String> TRANSFER_QUEUE =
                            new LinkedTransferQueue<>();

    public static void main(String[] args) throws InterruptedException {
        System.out.println(TRANSFER_QUEUE.add("one"));
        System.out.println(TRANSFER_QUEUE.add("two"));
    }
}
```

执行结果如下：

```
true
true
```

观察源代码,会发现底层调用 xfer(e,true,ASYNC,0L)方法,默认返回值为 true,如图 7-26 所示。

```
public boolean add(E e) {
    xfer(e, haveData: true, ASYNC, nanos: 0L);   ← 生产数据时
    return true;
}
```

图 7-26　add 方法源代码

2. offer(E e)

添加数据,接收泛型入参,作为新加数据。由于此类是无界队列,所以此方法的正常返回值为 true,代码如下：

```java
//第7章/three/LinkedTransferQueueTest.java
public class LinkedTransferQueueTest {

    private static final LinkedTransferQueue<String> TRANSFER_QUEUE =
                            new LinkedTransferQueue<>();
```

```java
public static void main(String[] args) throws InterruptedException {
    System.out.println(TRANSFER_QUEUE.offer("one"));
    System.out.println(TRANSFER_QUEUE.offer("two"));
}
}
```

执行结果如下：

```
true
true
```

观察源代码，会发现底层调用 xfer(e,true,ASYNC,0L)方法，默认返回值为 true，如图 7-27 所示。

3. put(E e)

添加数据，接收泛型入参，作为新加数据。由于此类是无界队列，所以此方法不存在阻塞状态，代码如下：

```java
//第 7 章/three/LinkedTransferQueueTest.java
public class LinkedTransferQueueTest {

    private static final LinkedTransferQueue<String> TRANSFER_QUEUE =
                                        new LinkedTransferQueue<>();

    public static void main(String[] args) throws InterruptedException {
        TRANSFER_QUEUE.put("one");
        TRANSFER_QUEUE.put("two");
    }
}
```

观察源代码，会发现底层调用 xfer(e,true,ASYNC,0L)方法，如图 7-28 所示。

```
public boolean offer(E e) {
    xfer(e, haveData: true, ASYNC, nanos: 0L);  ← 生产数据时
    return true;
}
```

```
public void put(E e) {
    xfer(e, haveData: true, ASYNC, nanos: 0L);  ← 生产数据时
}
```

图 7-27　offer 方法源代码　　　　　图 7-28　put 方法源代码

4. poll()

检索并删除此队列的头部数据。如果检索成功，则返回删除的数据，如果检索失败，则返回 null，代码如下：

```java
//第 7 章/three/LinkedTransferQueueTest.java
public class LinkedTransferQueueTest {

    private static final LinkedTransferQueue<String> TRANSFER_QUEUE =
                                        new LinkedTransferQueue<>();
```

```java
public static void main(String[] args) throws InterruptedException {
    TRANSFER_QUEUE.put("one");
    TRANSFER_QUEUE.put("two");
    System.out.println(TRANSFER_QUEUE.poll());
    System.out.println(TRANSFER_QUEUE.poll());
    System.out.println(TRANSFER_QUEUE.poll());
}
}
```

执行结果如下：

```
one
two
null
```

观察源代码，会发现底层调用 xfer(null,false,NOW,0L)方法，如图 7-29 所示。

```
public E poll() {
    return xfer( e: null,  haveData: false,  NOW,  nanos: 0L);
}
```
消费数据时

图 7-29　poll 方法源代码

5. tryTransfer(E e)

添加数据，接收泛型入参，作为新加数据。此方法需要立即匹配到当前队列中的消费者，如果匹配成功，则返回值为 true，否则返回值为 false，不会把数据存入队列中，代码如下：

```java
//第 7 章/three/LinkedTransferQueueTest.java
public class LinkedTransferQueueTest {

    private static final LinkedTransferQueue<String> TRANSFER_QUEUE =
                                        new LinkedTransferQueue<>();

    public static void main(String[] args) throws InterruptedException {
        System.out.println(TRANSFER_QUEUE.tryTransfer("one"));
        System.out.println(TRANSFER_QUEUE.tryTransfer("two"));
        System.out.println(TRANSFER_QUEUE.poll());
        System.out.println(TRANSFER_QUEUE.poll());
    }
}
```

执行结果如下：

```
false
false
null
null
```

观察源代码，会发现底层调用 xfer(e,true,NOW,0L)方法，如图 7-30 所示。

```
public boolean tryTransfer(E e) {
    return xfer(e, haveData: true, NOW, nanos: 0L) == null;
}
```
生产数据时

图 7-30　tryTransfer 方法源代码

6. take()

检索并删除此队列的头部数据。如果检索成功,则返回删除的数据,如果检索失败,则阻塞等待,代码如下：

```java
//第 7 章/three/LinkedTransferQueueTest.java
public class LinkedTransferQueueTest {

    private static final LinkedTransferQueue<String> TRANSFER_QUEUE =
                                new LinkedTransferQueue<>();

    public static void main(String[] args) throws InterruptedException {
        TransferRun transferRun = new TransferRun();
        new Thread(transferRun).start();
        System.out.println(Thread.currentThread().getName()
                    + ":" + TRANSFER_QUEUE.take());     //阻塞等待消费数据
    }

    static final class TransferRun implements Runnable{
        @Override
        public void run() {
            try {
                Thread.sleep(2000);
                TRANSFER_QUEUE.tryTransfer("one");           //生产数据
            } catch (InterruptedException e) {
                e.printStackTrace();
            }
        }
    }
}
```

执行结果如下：

```
main:one
```

观察源代码,会发现底层调用 xfer(null,false,SYNC,0L)方法,如图 7-31 所示。

```
public E take() throws InterruptedException {
    E e = xfer(e: null, haveData: false, SYNC, nanos: 0L);
    if (e != null)
        return e;
    Thread.interrupted();
    throw new InterruptedException();
}
```
方法类型

图 7-31　take 方法源代码

阻塞等待在 awaitMatch(s, p, e, (how == TIMED), nanos) 方法中实现,如图 7-32 所示。

```java
private E awaitMatch(Node s, Node pred, E e, boolean timed, long nanos) {
    final boolean isData = s.isData;
    final long deadline = timed ? System.nanoTime() + nanos : 0L;
    final Thread w = Thread.currentThread();// 当前执行线程对象
    int stat = -1;
    Object item;
    while ((item = s.item) == e) {
        if (needSweep)                      // 帮助清理
            sweep();
        else if ((timed && nanos <= 0L) || w.isInterrupted()) { //最大时间的阻塞等待已过
            if (s.casItem(e, (e == null) ? s : null)) {
                unsplice(pred, s);    // 取消
                return e;
            }
        }
        else if (stat <= 0) {
            if (pred != null && pred.next == s) {
                if (stat < 0 &&
                    (pred.isData != isData || pred.isMatched())) {
                    stat = 0;
                    Thread.yield();
                }
                else {
                    stat = 1;
                    s.waiter = w;     // 设置当前阻塞线程
                }
            }
        }
        else if ((item = s.item) != e)
            break;                          // 重新检查
        else if (!timed) {
            LockSupport.setCurrentBlocker(this);
            try {
                ForkJoinPool.managedBlock(s);// 没有时间的阻塞等待
            } catch (InterruptedException cannotHappen) { }
            LockSupport.setCurrentBlocker(null);
        }
        else {
            nanos = deadline - System.nanoTime();
            if (nanos > SPIN_FOR_TIMEOUT_THRESHOLD)
                LockSupport.parkNanos(this, nanos);// 最大时间的阻塞等待
        }
    }
    if (stat == 1)
        WAITER.set(s, null);
    if (!isData)
        ITEM.set(s, s);                     // 自链接以避免垃圾
    return (E) item;
}
```

图 7-32　awaitMatch 方法源代码

7. transfer(E e)

添加数据,接收泛型入参,作为新加数据。此方法会匹配当前队列中的消费者,如果匹配成功,则正常消费结束,如果匹配失败,则阻塞等待,代码如下:

```java
//第 7 章/three/LinkedTransferQueueTest.java
public class LinkedTransferQueueTest {

    private static final LinkedTransferQueue<String> TRANSFER_QUEUE =
                                        new LinkedTransferQueue<>();

    public static void main(String[] args) throws InterruptedException {
        TransferRun transferRun = new TransferRun();
        new Thread(transferRun).start();
        System.out.println(Thread.currentThread().getName()
                        + ":" + TRANSFER_QUEUE.take()); //阻塞等待消费数据
    }

    static final class TransferRun implements Runnable{
        @Override
        public void run() {
            try {
                Thread.sleep(2000);
                TRANSFER_QUEUE.transfer("one");        //生产数据,当没有消费者时阻塞等待
            } catch (InterruptedException e) {
                e.printStackTrace();
            }
        }
    }
}
```

执行结果如下:

```
main:one
```

观察源代码,会发现底层调用 xfer(e,true,SYNC,0L)方法,如图 7-33 所示。

```java
public void transfer(E e) throws InterruptedException {
    if (xfer(e, true, SYNC, 0L) != null) {
        Thread.interrupted();
        throw new InterruptedException();
    }
}
```

图 7-33　transfer 方法源代码

8. poll(long timeout,TimeUnit unit)

检索并删除此队列的头部数据。接收 long 入参,作为最大等待时间,接收 TimeUnit 入参,作为时间单位。如果检索成功,则返回删除的数据,如果检索失败,则给定最大等待时间段,在此时间段内如果检索成功,则返回删除的数据,否则返回 null,代码如下:

```java
//第7章/three/LinkedTransferQueueTest.java
public class LinkedTransferQueueTest {

    private static final LinkedTransferQueue< String > TRANSFER_QUEUE =
                                    new LinkedTransferQueue<>();

    public static void main(String[] args) throws InterruptedException {
        TRANSFER_QUEUE.add("one");
        System.out.println(TRANSFER_QUEUE.poll(2,TimeUnit.SECONDS));
        System.out.println(TRANSFER_QUEUE.poll(2,TimeUnit.SECONDS));
    }

}
```

执行结果如下：

```
one
null
```

9. tryTransfer(E e,long timeout,TimeUnit unit)

添加数据，接收泛型入参，作为新加数据，接收 long 入参，作为最大等待时间，接收 TimeUnit 入参，作为时间单位。此方法需要匹配到当前队列中的消费者，如果在给定最大等待时间内匹配成功，则返回值为 true，否则返回值为 false，不会把数据存入队列中，代码如下：

```java
//第7章/three/LinkedTransferQueueTest.java
public class LinkedTransferQueueTest {

    private static final LinkedTransferQueue< String > TRANSFER_QUEUE =
                                    new LinkedTransferQueue<>();

    public static void main(String[] args) throws InterruptedException {
        System.out.println(TRANSFER_QUEUE.tryTransfer("one",
                           2, TimeUnit.SECONDS));
        System.out.println(TRANSFER_QUEUE.tryTransfer("two",
                           2, TimeUnit.SECONDS));
        System.out.println(TRANSFER_QUEUE.poll());
    }

}
```

执行结果如下：

```
false
false
null
```

观察源代码，会发现底层调用 xfer(e,true,TIMED,unit.toNanos(timeout))方法，如

图 7-34 所示。

```
public boolean tryTransfer(E e, long timeout, TimeUnit unit)
    throws InterruptedException {
    if (xfer(e, haveData: true, TIMED, unit.toNanos(timeout)) == null)    ← 方法类型
        return true;
    if (!Thread.interrupted())
        return false;
    throw new InterruptedException();
}
```

图 7-34　tryTransfer 源代码

10. getWaitingConsumerCount()

返回等待消费者数量的估计值，代码如下：

```java
public class LinkedTransferQueueTest {

    private static final LinkedTransferQueue<String> TRANSFER_QUEUE =
                        new LinkedTransferQueue<>();

    public static void main(String[] args) throws InterruptedException {
        TransferRun transferRun = new TransferRun();
        new Thread(transferRun).start();
        System.out.println(Thread.currentThread().getName()
                        + ":" + TRANSFER_QUEUE.take());
    }

    static final class TransferRun implements Runnable{
        @Override
        public void run() {
            try {
                Thread.sleep(2000);
                System.out.println(TRANSFER_QUEUE
                                .getWaitingConsumerCount());
                TRANSFER_QUEUE.tryTransfer("one",1,TimeUnit.SECONDS);
            } catch (InterruptedException e) {
                e.printStackTrace();
            }
        }
    }
}
```

执行结果如下：

```
1
main:one
```

11. contains(Object o)

如果此队列包含指定的元素，则返回 true，否则返回 false。接收 Object 入参，作为比较

的元素。

12. remainingCapacity()

返回此队列在理想情况下可以接受的元素数量,由于此类是无界队列,所以此方法会固定返回 Integer.MAX_VALUE,如图 7-35 所示。

```
public int remainingCapacity() {
    return Integer.MAX_VALUE;
}
```

图 7-35 remainingCapacity 方法源代码

13. size()

返回此队列中的已添加元素的数量,代码如下:

```
public class LinkedTransferQueueTest {

    private static final LinkedTransferQueue<String> TRANSFER_QUEUE =
                                    new LinkedTransferQueue<>();

    public static void main(String[] args) throws InterruptedException {
        TRANSFER_QUEUE.add("one");
        TRANSFER_QUEUE.put("one");
        TRANSFER_QUEUE.offer("one");
        System.out.println(TRANSFER_QUEUE.size());
    }

}
```

执行结果如下:

```
3
```

7.4 SynchronousQueue

一个无容量阻塞队列,其中每个插入操作都必须等待另一个线程的相应删除操作,反之亦然。底层基于 VarHandle 实现原子操作,可选的公平策略。

26min

7.4.1 构造器

SynchronousQueue 构造器见表 7-5。

表 7-5 SynchronousQueue 构造器

构造器	描述
SynchronousQueue()	构造新的对象,默认为无参构造器
SynchronousQueue(boolean fair)	构造新的对象,指定公平策略

非公平策略底层使用 TransferStack 类,公平策略底层使用 TransferQueue 类,如图 7-36 所示。

非公平策略底层结构类似栈,代码如下:

```
public SynchronousQueue(boolean fair) {
    transferer = fair ? new TransferQueue<E>() : new TransferStack<E>();
}
```

图 7-36 SynchronousQueue 构造器源代码

```java
//第 7 章/four/SynchronousQueueTest.java
public class SynchronousQueueTest {

    private static final SynchronousQueue<String> SYNCHRONOUS_QUEUE =
                            new SynchronousQueue<>(false);

    public static void main(String[] args) throws InterruptedException {
        SynchronousQueueRun synchronousQueueRun = new SynchronousQueueRun();
        new Thread(synchronousQueueRun,"one").start();
        Thread.sleep(200);
        new Thread(synchronousQueueRun,"two").start();
        Thread.sleep(200);
        new Thread(synchronousQueueRun,"three").start();
        Thread.sleep(200);

        System.out.println(SYNCHRONOUS_QUEUE.take());
        System.out.println(SYNCHRONOUS_QUEUE.poll());
        System.out.println(SYNCHRONOUS_QUEUE.take());
    }

    static final class SynchronousQueueRun implements Runnable{
        @Override
        public void run() {
            try {
                SYNCHRONOUS_QUEUE.put(Thread.currentThread().getName());
            } catch (Exception e) {
                e.printStackTrace();
            }
        }
    }
}
```

执行结果如下:

```
three
two
one
```

公平策略底层结构 FIFO(先进先出)队列,代码如下:

```java
//第 7 章/four/SynchronousQueueTest.java
public class SynchronousQueueTest {

    private static final SynchronousQueue<String> SYNCHRONOUS_QUEUE =
                            new SynchronousQueue<>(true);
```

```java
public static void main(String[] args) throws InterruptedException {
    SynchronousQueueRun synchronousQueueRun = new SynchronousQueueRun();
    new Thread(synchronousQueueRun,"one").start();
    Thread.sleep(200);
    new Thread(synchronousQueueRun,"two").start();
    Thread.sleep(200);
    new Thread(synchronousQueueRun,"three").start();
    Thread.sleep(200);

    System.out.println(SYNCHRONOUS_QUEUE.take());
    System.out.println(SYNCHRONOUS_QUEUE.poll());
    System.out.println(SYNCHRONOUS_QUEUE.take());
}

static final class SynchronousQueueRun implements Runnable{
    @Override
    public void run() {
        try {
            SYNCHRONOUS_QUEUE.put(Thread.currentThread().getName());
        } catch (Exception e) {
            e.printStackTrace();
        }
    }
}
```

执行结果如下:

```
one
two
three
```

7.4.2 常用方法

1. add(E e)

添加数据,接收泛型入参,作为新加数据。由于此类是无容量队列,所以此方法需要立即匹配到当前队列中的消费节点,如果匹配上,则返回值为 true,否则将引发 IllegalStateException 异常,代码如下:

```java
//第7章/four/SynchronousQueueTest.java
public class SynchronousQueueTest {

    private static final SynchronousQueue<String> SYNCHRONOUS_QUEUE =
                            new SynchronousQueue<>(true);

    public static void main(String[] args) throws InterruptedException {
        SynchronousQueueRun synchronousQueueRun = new SynchronousQueueRun();
        new Thread(synchronousQueueRun).start();
```

```java
            Thread.sleep(200);
            System.out.println(SYNCHRONOUS_QUEUE.add("one"));
        }

        static final class SynchronousQueueRun implements Runnable{
            @Override
            public void run() {
                try {
                    System.out.println(SYNCHRONOUS_QUEUE.take());
                                                    //消费数据,阻塞等待生产者
                } catch (Exception e) {
                    e.printStackTrace();
                }
            }
        }
    }
```

执行结果如下:

```
true
one
```

当添加数据匹配不上消费节点时,代码如下:

```java
//第7章/four/SynchronousQueueTest.java
public class SynchronousQueueTest {

    private static final SynchronousQueue<String> SYNCHRONOUS_QUEUE =
                                new SynchronousQueue<>(true);

    public static void main(String[] args) throws InterruptedException {
        System.out.println(SYNCHRONOUS_QUEUE.add("one"));
    }
}
```

执行结果如下:

```
Exception in thread "main" java.lang.IllegalStateException: Queue full
    at java.base/java.util.AbstractQueue.add(AbstractQueue.java:98)
    at cn.kungreat.book.seven.four.SynchronousQueueTest.main
(SynchronousQueueTest.java:11)
```

2. offer(E e)

添加数据,接收泛型入参,作为新加数据。由于此类是无容量队列,所以此方法需要立即匹配到当前队列中的消费节点,如果匹配上,则返回值为 true,否则返回值为 false,代码如下:

```java
//第7章/four/SynchronousQueueTest.java
public class SynchronousQueueTest {

    private static final SynchronousQueue<String> SYNCHRONOUS_QUEUE =
                                    new SynchronousQueue<>(true);

    public static void main(String[] args) throws InterruptedException {
        SynchronousQueueRun synchronousQueueRun = new SynchronousQueueRun();
        new Thread(synchronousQueueRun).start();
        Thread.sleep(200);
        System.out.println(SYNCHRONOUS_QUEUE.offer("one"));
    }

    static final class SynchronousQueueRun implements Runnable{
        @Override
        public void run() {
            try {
                System.out.println(SYNCHRONOUS_QUEUE.take());
                                        //消费数据,阻塞等待生产者
            } catch (Exception e) {
                e.printStackTrace();
            }
        }
    }
}
```

执行结果如下:

```
true
one
```

3. remove()

检索并删除此队列的头部数据。由于此类是无容量队列,所以此方法需要立即匹配到当前队列中的生产节点,如果匹配上,则返回生产的数据,否则将引发 NoSuchElementException 异常,代码如下:

```java
//第7章/four/SynchronousQueueTest.java
public class SynchronousQueueTest {

    private static final SynchronousQueue<String> SYNCHRONOUS_QUEUE =
                                    new SynchronousQueue<>(true);

    public static void main(String[] args) throws InterruptedException {
        SynchronousQueueRun synchronousQueueRun = new SynchronousQueueRun();
        new Thread(synchronousQueueRun).start();
        Thread.sleep(200);
        System.out.println(SYNCHRONOUS_QUEUE.remove());
    }
```

```java
        static final class SynchronousQueueRun implements Runnable{
            @Override
            public void run() {
                try {
                    SYNCHRONOUS_QUEUE.put("one"); //生产数据,阻塞等待消费者
                } catch (Exception e) {
                    e.printStackTrace();
                }
            }
        }
```

执行结果如下:

```
one
```

当匹配不上生产节点时,代码如下:

```java
//第 7 章/four/SynchronousQueueTest.java
public class SynchronousQueueTest {

    private static final SynchronousQueue<String> SYNCHRONOUS_QUEUE =
                                    new SynchronousQueue<>(true);

    public static void main(String[] args) throws InterruptedException {
        System.out.println(SYNCHRONOUS_QUEUE.remove());
    }
}
```

执行结果如下:

```
Exception in thread "main" java.util.NoSuchElementException
    at java.base/java.util.AbstractQueue.remove(AbstractQueue.java:117)
    at cn.kungreat.book.seven.four.SynchronousQueueTest.main
(SynchronousQueueTest.java:11)
```

此方法底层调用的是 poll()方法,如图 7-37 所示。

```java
public E remove() {
    E x = poll();
    if (x != null)
        return x;
    else
        throw new NoSuchElementException();
}
```

图 7-37 remove 方法源代码

4. poll()

检索并删除此队列的头部数据。由于此类是无容量队列,所以此方法需要立即匹配到

当前队列中的生产节点,如果匹配上,则返回生产的数据,否则返回 null,代码如下:

```java
//第 7 章/four/SynchronousQueueTest.java
public class SynchronousQueueTest {

    private static final SynchronousQueue<String> SYNCHRONOUS_QUEUE =
                                  new SynchronousQueue<>(true);

    public static void main(String[] args) throws InterruptedException {
        SynchronousQueueRun synchronousQueueRun = new SynchronousQueueRun();
        new Thread(synchronousQueueRun).start();
        Thread.sleep(200);
        System.out.println(SYNCHRONOUS_QUEUE.poll());
    }

    static final class SynchronousQueueRun implements Runnable{
        @Override
        public void run() {
            try {
                SYNCHRONOUS_QUEUE.put("one"); //生产数据,阻塞等待消费者
            } catch (Exception e) {
                e.printStackTrace();
            }
        }
    }
}
```

执行结果如下:

```
one
```

5. element()

检索但不删除此队列的头部数据。由于此类是无容量队列,所以此方法将直接引发 NoSuchElementException 异常,代码如下:

```java
//第 7 章/four/SynchronousQueueTest.java
public class SynchronousQueueTest {

    private static final SynchronousQueue<String> SYNCHRONOUS_QUEUE =
                                  new SynchronousQueue<>(true);

    public static void main(String[] args) throws InterruptedException {
        SynchronousQueueRun synchronousQueueRun = new SynchronousQueueRun();
        new Thread(synchronousQueueRun).start();
        Thread.sleep(200);
        System.out.println(SYNCHRONOUS_QUEUE.element());
    }

    static final class SynchronousQueueRun implements Runnable{
        @Override
```

```java
    public void run() {
        try {
            SYNCHRONOUS_QUEUE.put("one"); //生产数据,阻塞等待消费者
        } catch (Exception e) {
            e.printStackTrace();
        }
    }
}
```

执行结果如下:

```
Exception in thread "main" java.util.NoSuchElementException
    at java.base/java.util.AbstractQueue.element(AbstractQueue.java:136)
    at cn.kungreat.book.seven.four.SynchronousQueueTest.main
(SynchronousQueueTest.java:14)
```

注意：此程序并没有结束,因为生产数据的节点还在阻塞等待消费者。

```java
public E peek() {
    return null;
}
```

图 7-38 peek 方法源代码

6. peek()

检索但不删除此队列的头部数据。由于此类是无容量队列,所以此方法将直接返回空,如图 7-38 所示。

7. put(E e)

添加数据,接收泛型入参,作为新加数据。由于此类是无容量队列,所以此方法将会一直阻塞等待匹配消费者,匹配成功后正常结束方法,代码如下:

```java
//第 7 章/four/SynchronousQueueTest.java
public class SynchronousQueueTest {

    private static final SynchronousQueue<String> SYNCHRONOUS_QUEUE =
                        new SynchronousQueue<>(true);

    public static void main(String[] args) throws InterruptedException {
        SynchronousQueueRun synchronousQueueRun = new SynchronousQueueRun();
        new Thread(synchronousQueueRun).start();
        Thread.sleep(200);
        System.out.println(SYNCHRONOUS_QUEUE.take());
                //消费数据,阻塞等待生产者
    }

    static final class SynchronousQueueRun implements Runnable{
        @Override
        public void run() {
            try {
                SYNCHRONOUS_QUEUE.put("one"); //生产数据,阻塞等待消费者
            } catch (Exception e) {
```

```
            e.printStackTrace();
        }
    }
}
```

执行结果如下:

```
one
```

8. take()

检索并删除此队列的头部数据。由于此类是无容量队列,所以此方法将会一直阻塞等待匹配生产者,匹配成功后将返回生产者数据。

9. offer(E e, long timeout, TimeUnit unit)

添加数据,接收泛型入参,作为新加数据,接收 long 入参,作为最大等待时间,接收 TimeUnit 入参,作为时间单位。由于此类是无容量队列,所以此方法需要匹配到当前队列中的消费节点,如果在最大等待给定时间内匹配上,则返回值为 true,否则将返回值为 false,代码如下:

```
//第 7 章/four/SynchronousQueueTest.java
public class SynchronousQueueTest {

    private static final SynchronousQueue<String> SYNCHRONOUS_QUEUE =
                             new SynchronousQueue<>(true);

    public static void main(String[] args) throws InterruptedException {
        SynchronousQueueRun synchronousQueueRun = new SynchronousQueueRun();
        new Thread(synchronousQueueRun).start();
        Thread.sleep(1000);
        System.out.println(SYNCHRONOUS_QUEUE.take());
    }

    static final class SynchronousQueueRun implements Runnable{
        @Override
        public void run() {
            try {
                System.out.println("offer:" +
                    SYNCHRONOUS_QUEUE.offer("one", 2, TimeUnit.SECONDS));
            } catch (Exception e) {
                e.printStackTrace();
            }
        }
    }
}
```

执行结果如下:

```
one
offer:true
```

10. poll(long timeout,TimeUnit unit)

检索并删除此队列的头部数据。接收 long 入参,作为最大等待时间,接收 TimeUnit 入参,作为时间单位。由于此类是无容量队列,所以此方法需要匹配到当前队列中的生产节点,如果在最大等待给定时间内匹配上,则返回生产的数据,否则返回 null,代码如下:

```java
//第7章/four/SynchronousQueueTest.java
public class SynchronousQueueTest {

    private static final SynchronousQueue<String> SYNCHRONOUS_QUEUE =
                                new SynchronousQueue<>(true);

    public static void main(String[] args) throws InterruptedException {
        System.out.println(SYNCHRONOUS_QUEUE.poll(2,TimeUnit.SECONDS));
    }
}
```

执行结果如下:

```
null
```

11. clear()

由于此类是无容量队列,所以此方法是空实现,如图 7-39 所示。

```java
public void clear() {
}
```

图 7-39 clear 方法源代码

12. contains(Object o)

由于此类是无容量队列,所以此方法的固定返回值为 false,如图 7-40 所示。

13. isEmpty()

由于此类是无容量队列,所以此方法的固定返回值为 true,如图 7-41 所示。

```java
public boolean contains(Object o) {
    return false;
}
```

图 7-40 contains 方法源代码

```java
public boolean isEmpty() {
    return true;
}
```

图 7-41 isEmpty 方法源代码

14. remainingCapacity()

返回此队列在理想情况下还可以接受的元素数量,由于此类是无容量队列,所以此方法的固定返回值为 0,如图 7-42 所示。

15. size()

返回此队列中的已添加元素的数量,由于此类是无容量队列,所以此方法的固定返回值为 0,如图 7-43 所示。

```
public int remainingCapacity() {
    return 0;
}
```

图 7-42 remainingCapacity 方法源代码

```
public int size() {
    return 0;
}
```

图 7-43 size 方法源代码

7.5 DelayQueue

一个无界阻塞队列，底层存储数据使用 PriorityQueue 类，底层防并发使用 ReentrantLock 锁，泛型实现类需要去实现排序方法和延迟方法。

31min

7.5.1 构造器

DelayQueue 构造器见表 7-6。

表 7-6 DelayQueue 构造器

构造器	描述
DelayQueue()	构造新的对象，默认为无参构造器
DelayQueue(Collection<? extends E> c)	构造新的对象，指定初始数据

7.5.2 常用方法

1. add(E e)

添加数据，接收泛型入参，作为新加数据。由于此类是无界队列，所以此方法的正常返回值为 true，代码如下：

```java
//第7章/five/DelayQueueTest.java
public class DelayQueueTest {

    private static final DelayQueue<DelayedImpl> DELAY_QUEUE =
                                                    new DelayQueue<>();

    public static void main(String[] args) throws Exception {
        long currentTimeMillis = System.currentTimeMillis();
        DELAY_QUEUE.add(new DelayedImpl(20, currentTimeMillis + 3000));
        DELAY_QUEUE.add(new DelayedImpl(5, currentTimeMillis + 5000));
        DELAY_QUEUE.add(new DelayedImpl(3, currentTimeMillis + 10000));
        DELAY_QUEUE.add(new DelayedImpl(80, currentTimeMillis + 2000));
        DELAY_QUEUE.add(new DelayedImpl(160, currentTimeMillis + 2000));
        System.out.println(DELAY_QUEUE.take());
        System.out.println(DELAY_QUEUE.take());
        System.out.println(DELAY_QUEUE.take());
        System.out.println(DELAY_QUEUE.take());
        System.out.println(DELAY_QUEUE.take());
    }
}
```

```java
        static final class DelayedImpl implements Delayed {
            private Integer comp;
            private Long delayTime;

            public DelayedImpl(Integer comp, Long delayTime) {
                this.comp = comp;
                this.delayTime = delayTime;
            }

            @Override
            public long getDelay(TimeUnit unit) {
                return delayTime - System.currentTimeMillis();
            }

            @Override
            public int compareTo(Delayed o) {
                DelayedImpl delayed = (DelayedImpl) o;
                return comp - delayed.comp;
            }

            @Override
            public String toString() {
                return "DelayedImpl{" +
                        "comp=" + comp +
                        '}';
            }
        }
    }
```

执行结果如下：

```
DelayedImpl{comp=3}
DelayedImpl{comp=5}
DelayedImpl{comp=20}
DelayedImpl{comp=80}
DelayedImpl{comp=160}
```

注意：数据的生产和消费都会有排序效果，数据的消费具有延迟效果。

2. offer(E e)

添加数据，接收泛型入参，作为新加数据。由于此类是无界队列，所以此方法的正常返回值为 true，如图 7-44 所示。

3. put(E e)

添加数据，接收泛型入参，作为新加数据。由于此类是无界队列，所以此方法不存在阻塞状态。

```
public boolean offer(E e) {
    final ReentrantLock lock = this.lock;
    lock.lock();
    try {
        q.offer(e);
        if (q.peek() == e) {
            leader = null;
            available.signal();
        }
        return true;
    } finally {
        lock.unlock();
    }
}
```

图 7-44　offer 方法源代码

4. remove()

检索并删除此队列的头部数据。如果检索成功，则返回删除的数据，如果此队列为空，它将引发 NoSuchElementException 异常，代码如下：

```
//第 7 章/five/DelayQueueTest.java
public class DelayQueueTest {

    private static final DelayQueue<DelayedImpl> DELAY_QUEUE =
                                    new DelayQueue<>();

    public static void main(String[] args) throws Exception {
        long currentTimeMillis = System.currentTimeMillis();
        DELAY_QUEUE.put(new DelayedImpl(20,currentTimeMillis));
        System.out.println(DELAY_QUEUE.remove());
    }

    static final class DelayedImpl implements Delayed {

        private Integer comp;
        private Long delayTime;

        public DelayedImpl(Integer comp, Long delayTime) {
            this.comp = comp;
            this.delayTime = delayTime;
        }

        @Override
        public long getDelay(TimeUnit unit) {
            return delayTime - System.currentTimeMillis();
        }

        @Override
```

```
        public int compareTo(Delayed o) {
            DelayedImpl delayed = (DelayedImpl) o;
            return comp - delayed.comp;
        }

        @Override
        public String toString() {
            return "DelayedImpl{" +
                    "comp = " + comp +
                    '}';
        }
    }
}
```

执行结果如下：

```
DelayedImpl{comp = 20}
```

5. poll()

检索并删除此队列的头部数据。如果检索成功，则返回删除的数据，如果检索失败，则返回 null，代码如下：

```
//第 7 章/five/DelayQueueTest.java
public class DelayQueueTest {

    private static final DelayQueue<DelayedImpl> DELAY_QUEUE =
                                        new DelayQueue<>();

    public static void main(String[] args) throws Exception {
        long currentTimeMillis = System.currentTimeMillis();
        DELAY_QUEUE.add(new DelayedImpl(5, currentTimeMillis + 5000));
        DELAY_QUEUE.add(new DelayedImpl(3, currentTimeMillis + 10000));
        DELAY_QUEUE.add(new DelayedImpl(80, currentTimeMillis + 2000));
        DELAY_QUEUE.add(new DelayedImpl(160, currentTimeMillis + 2000));
        System.out.println(DELAY_QUEUE.poll());
        System.out.println(DELAY_QUEUE.poll());
    }

    static final class DelayedImpl implements Delayed {

        private Integer comp;
        private Long delayTime;

        public DelayedImpl(Integer comp, Long delayTime) {
            this.comp = comp;
            this.delayTime = delayTime;
        }

        @Override
```

```java
    public long getDelay(TimeUnit unit) {
        return delayTime - System.currentTimeMillis();
    }

    @Override
    public int compareTo(Delayed o) {
        DelayedImpl delayed = (DelayedImpl) o;
        return comp - delayed.comp;
    }

    @Override
    public String toString() {
        return "DelayedImpl{" +
                "comp=" + comp +
                '}';
    }
  }
}
```

执行结果如下：

```
null
null
```

注意：数据的消费具有延迟效果。

6. take()

检索并删除此队列的头部数据。如果检索成功，则返回删除的数据；如果检索失败，则阻塞等待，代码如下：

```java
//第 7 章/five/DelayQueueTest.java
public class DelayQueueTest {

    private static final DelayQueue<DelayedImpl> DELAY_QUEUE =
                                            new DelayQueue<>();

    public static void main(String[] args) throws Exception {
        long currentTimeMillis = System.currentTimeMillis();
        DELAY_QUEUE.add(new DelayedImpl(5, currentTimeMillis + 1000));
        DELAY_QUEUE.add(new DelayedImpl(3, currentTimeMillis + 2000));
        DELAY_QUEUE.add(new DelayedImpl(80, currentTimeMillis + 500));
        DELAY_QUEUE.add(new DelayedImpl(160, currentTimeMillis + 300));
        System.out.println(DELAY_QUEUE.take());
        System.out.println(DELAY_QUEUE.take());
        System.out.println(DELAY_QUEUE.take());
        System.out.println(DELAY_QUEUE.take());
    }
}
```

```java
    static final class DelayedImpl implements Delayed {
        private Integer comp;
        private Long delayTime;

        public DelayedImpl(Integer comp, Long delayTime) {
            this.comp = comp;
            this.delayTime = delayTime;
        }

        @Override
        public long getDelay(TimeUnit unit) {
            return delayTime - System.currentTimeMillis();
        }

        @Override
        public int compareTo(Delayed o) {
            DelayedImpl delayed = (DelayedImpl) o;
            return comp - delayed.comp;
        }

        @Override
        public String toString() {
            return "DelayedImpl{" +
                    "comp = " + comp +
                    '}';
        }
    }
}
```

执行结果如下：

```
DelayedImpl{comp = 3}
DelayedImpl{comp = 5}
DelayedImpl{comp = 80}
DelayedImpl{comp = 160}
```

7. poll(long timeout，TimeUnit unit)

检索并删除此队列的头部数据。接收 long 入参，作为最大等待时间，接收 TimeUnit 入参，作为时间单位。如果在最大等待给定时间段内检索成功，则返回删除的数据，否则返回 null，代码如下：

```java
//第 7 章/five/DelayQueueTest.java
public class DelayQueueTest {

    private static final DelayQueue<DelayedImpl> DELAY_QUEUE =
                                            new DelayQueue<>();
```

```java
    public static void main(String[] args) throws Exception {
        long currentTimeMillis = System.currentTimeMillis();
        DELAY_QUEUE.add(new DelayedImpl(5,currentTimeMillis + 1000));
        DELAY_QUEUE.add(new DelayedImpl(3,currentTimeMillis + 2000));
        DELAY_QUEUE.add(new DelayedImpl(80,currentTimeMillis + 500));
        DELAY_QUEUE.add(new DelayedImpl(160,currentTimeMillis + 300));
        System.out.println(DELAY_QUEUE.poll(1,TimeUnit.SECONDS));
        System.out.println(DELAY_QUEUE.poll(2,TimeUnit.SECONDS));
        System.out.println(DELAY_QUEUE.poll(2,TimeUnit.SECONDS));
        System.out.println(DELAY_QUEUE.poll(2,TimeUnit.SECONDS));
    }

    static final class DelayedImpl implements Delayed {

        private Integer comp;
        private Long delayTime;

        public DelayedImpl(Integer comp, Long delayTime) {
            this.comp = comp;
            this.delayTime = delayTime;
        }

        @Override
        public long getDelay(TimeUnit unit) {
            return delayTime - System.currentTimeMillis();
        }

        @Override
        public int compareTo(Delayed o) {
            DelayedImpl delayed = (DelayedImpl) o;
            return comp - delayed.comp;
        }

        @Override
        public String toString() {
            return "DelayedImpl{" +
                    "comp=" + comp +
                    '}';
        }
    }
}
```

执行结果如下：

```
null
DelayedImpl{comp=3}
DelayedImpl{comp=5}
DelayedImpl{comp=80}
```

8. offer(E e, long timeout, TimeUnit unit)

添加数据，接收泛型入参，作为新加数据。由于此类是无界队列，所以此方法的底层会直接调用 offer(E e) 方法,如图 7-45 所示。

```
public boolean offer(E e, long timeout, TimeUnit unit) {
    return offer(e);
}
```

图 7-45　offer 方法源代码

9. peek()

检索但不删除此队列的头部数据。如果成功,则返回检索的数据,如果此队列为空,则返回 null。

10. element()

检索但不删除此队列的头部数据。如果成功,则返回检索的数据,如果此队列为空,它将引发 NoSuchElementException 异常。

11. drainTo(Collection <? super E > c)

从此队列中删除所有可用元素,并将它们添加到给定集合中,然后返回添加的数量。接收 Collection 入参,作为给定集合,代码如下:

```java
//第7章/five/DelayQueueTest.java
public class DelayQueueTest {

    private static final DelayQueue < DelayedImpl > DELAY_QUEUE =
                                         new DelayQueue<>();

    public static void main(String[] args) throws Exception {
        long currentTimeMillis = System.currentTimeMillis();
        DELAY_QUEUE.offer(new DelayedImpl(5, currentTimeMillis + 1000));
        DELAY_QUEUE.add(new DelayedImpl(3, currentTimeMillis + 2000));
        DELAY_QUEUE.add(new DelayedImpl(80, currentTimeMillis + 500));
        DELAY_QUEUE.add(new DelayedImpl(160, currentTimeMillis + 300));
        List < DelayedImpl > list = new ArrayList<>();
        Thread.sleep(3000);
        System.out.println(DELAY_QUEUE.drainTo(list));
        System.out.println(list);
    }

    static final class DelayedImpl implements Delayed {

        private Integer comp;
        private Long delayTime;

        public DelayedImpl(Integer comp, Long delayTime) {
            this.comp = comp;
            this.delayTime = delayTime;
        }

        @Override
        public long getDelay(TimeUnit unit) {
```

```
            return delayTime - System.currentTimeMillis();
        }

        @Override
        public int compareTo(Delayed o) {
            DelayedImpl delayed = (DelayedImpl) o;
            return comp - delayed.comp;
        }

        @Override
        public String toString() {
            return "DelayedImpl{" +
                    "comp=" + comp +
                    '}';
        }
    }
}
```

执行结果如下：

```
4
[DelayedImpl{comp=3}, DelayedImpl{comp=5}, DelayedImpl{comp=80}, DelayedImpl{comp=160}]
```

12. remainingCapacity()

返回此队列在理想情况下还可以接受的元素数量，由于此类是无界队列，所以此方法固定返回 Integer.MAX_VALUE，如图 7-46 所示。

```
public int remainingCapacity() {
    return Integer.MAX_VALUE;
}
```

图 7-46　remainingCapacity 方法源代码

13. size()

返回此队列中已添加元素的数量，代码如下：

```
//第 7 章/five/DelayQueueTest.java
public class DelayQueueTest {

    private static final DelayQueue<DelayedImpl> DELAY_QUEUE =
                                        new DelayQueue<>();

    public static void main(String[] args) throws Exception {
        long currentTimeMillis = System.currentTimeMillis();
        DELAY_QUEUE.offer(new DelayedImpl(5,currentTimeMillis + 1000));
        DELAY_QUEUE.add(new DelayedImpl(3,currentTimeMillis + 2000));
        DELAY_QUEUE.add(new DelayedImpl(80,currentTimeMillis + 500));
        DELAY_QUEUE.add(new DelayedImpl(160,currentTimeMillis + 300));
        System.out.println(DELAY_QUEUE.size());
    }
```

```java
static final class DelayedImpl implements Delayed {
    private Integer comp;
    private Long delayTime;

    public DelayedImpl(Integer comp, Long delayTime) {
        this.comp = comp;
        this.delayTime = delayTime;
    }

    @Override
    public long getDelay(TimeUnit unit) {
        return delayTime - System.currentTimeMillis();
    }

    @Override
    public int compareTo(Delayed o) {
        DelayedImpl delayed = (DelayedImpl) o;
        return comp - delayed.comp;
    }

    @Override
    public String toString() {
        return "DelayedImpl{" +
            "comp=" + comp +
            '}';
    }
}
```

执行结果如下：

4

7.6 PriorityBlockingQueue

一个无界阻塞队列，数据的生产和消费都有排序效果，排序效果和 PriorityQueue 类相同，泛型类需要去实现排序方法或者在构造对象时指定公共排序 Comparator 实现类。底层防并发使用 ReentrantLock 锁。

7.6.1 构造器

PriorityBlockingQueue 构造器见表 7-7。

表 7-7 PriorityBlockingQueue 构造器

构造器	描述
PriorityBlockingQueue()	构造新的对象，默认为无参构造器
PriorityBlockingQueue(int initialCapacity)	构造新的对象，指定初始容量

续表

构 造 器	描 述
PriorityBlockingQueue(int initialCapacity, Comparator<? super E> comparator)	构造新的对象,指定初始容量,指定公共比较器
PriorityBlockingQueue(Collection<? extends E> c)	构造新的对象,指定初始数据

7.6.2 常用方法

1. add(E e)

添加数据,接收泛型入参,作为新加数据。由于此类是无界队列,所以此方法的正常返回值为 true,代码如下:

```java
//第7章/six/PriorityBlockingQueueTest.java
public class PriorityBlockingQueueTest {

    private static final PriorityBlockingQueue<PriorityImpl> QUEUE =
                        new PriorityBlockingQueue<>(16,
                            (o1, o2) -> o1.comp - o2.comp);

    public static void main(String[] args) throws Exception {
        System.out.println(QUEUE.add(new PriorityImpl(15)));
        System.out.println(QUEUE.add(new PriorityImpl(2)));
        System.out.println(QUEUE.add(new PriorityImpl(66)));
        System.out.println(QUEUE.add(new PriorityImpl(78)));

        System.out.println(QUEUE.take());
        System.out.println(QUEUE.take());
        System.out.println(QUEUE.take());
        System.out.println(QUEUE.take());
    }

    static final class PriorityImpl {

        private Integer comp;

        public PriorityImpl(Integer comp) {
            this.comp = comp;
        }

        @Override
        public String toString() {
            return "PriorityImpl{" +
                    "comp = " + comp +
                    '}';
        }
    }
}
```

执行结果如下：

```
true
true
true
true
PriorityImpl{comp = 2}
PriorityImpl{comp = 15}
PriorityImpl{comp = 66}
PriorityImpl{comp = 78}
```

2. offer(E e)

添加数据，接收泛型入参，作为新加数据。由于此类是无界队列，所以此方法的正常返回值为 true，如图 7-47 所示。

```java
public boolean offer(E e) {
    if (e == null)
        throw new NullPointerException();
    final ReentrantLock lock = this.lock;
    lock.lock();
    int n, cap;
    Object[] es;
    while ((n = size) >= (cap = (es = queue).length))
        tryGrow(es, cap);
    try {
        final Comparator<? super E> cmp;
        if ((cmp = comparator) == null)
            siftUpComparable(n, e, es);
        else
            siftUpUsingComparator(n, e, es, cmp);
        size = n + 1;
        notEmpty.signal();
    } finally {
        lock.unlock();
    }
    return true;
}
```

图 7-47　offer 方法源代码

3. put(E e)

添加数据，接收泛型入参，作为新加数据。由于此类是无界队列，所以此方法不存在阻塞状态。

4. remove()

检索并删除此队列的头部数据。如果检索成功，则返回删除的数据，如果此队列为空，它将引发 NoSuchElementException 异常，代码如下：

```java
//第7章/six/PriorityBlockingQueueTest.java
public class PriorityBlockingQueueTest {

    private static final PriorityBlockingQueue<PriorityImpl> QUEUE =
                                    new PriorityBlockingQueue<>(16,
                                        (o1, o2) -> o1.comp - o2.comp);

    public static void main(String[] args) throws Exception {
        System.out.println(QUEUE.offer(new PriorityImpl(15)));
        System.out.println(QUEUE.offer(new PriorityImpl(2)));
        System.out.println(QUEUE.add(new PriorityImpl(66)));
        System.out.println(QUEUE.add(new PriorityImpl(78)));

        System.out.println(QUEUE.remove());
        System.out.println(QUEUE.remove());
        System.out.println(QUEUE.remove());
        System.out.println(QUEUE.remove());
        System.out.println(QUEUE.remove());
    }

    static final class PriorityImpl {

        private Integer comp;

        public PriorityImpl(Integer comp) {
            this.comp = comp;
        }

        @Override
        public String toString() {
            return "PriorityImpl{" +
                    "comp=" + comp +
                    '}';
        }
    }
}
```

执行结果如下：

```
true
true
true
true
PriorityImpl{comp=2}
PriorityImpl{comp=15}
PriorityImpl{comp=66}
PriorityImpl{comp=78}
Exception in thread "main" java.util.NoSuchElementException
    at java.base/java.util.AbstractQueue.remove(AbstractQueue.java:117)
    at cn.kungreat.book.seven.six.PriorityBlockingQueueTest.main
(PriorityBlockingQueueTest.java:21)
```

5. poll()

检索并删除此队列的头部数据。如果检索成功,则返回删除的数据,如果检索失败,则返回 null。

6. take()

检索并删除此队列的头部数据。如果检索成功,则返回删除的数据,如果检索失败,则阻塞等待,代码如下:

```java
//第7章/six/PriorityBlockingQueueTest.java
public class PriorityBlockingQueueTest {

    private static final PriorityBlockingQueue<PriorityImpl> QUEUE =
                            new PriorityBlockingQueue<>(16,
                            (o1, o2) -> o1.comp - o2.comp);

    public static void main(String[] args) throws Exception {
        System.out.println(QUEUE.offer(new PriorityImpl(15)));
        System.out.println(QUEUE.offer(new PriorityImpl(2)));
        System.out.println(QUEUE.add(new PriorityImpl(66)));
        System.out.println(QUEUE.add(new PriorityImpl(78)));

        System.out.println(QUEUE.take());
        System.out.println(QUEUE.take());
        System.out.println(QUEUE.take());
        System.out.println(QUEUE.take());
        System.out.println(QUEUE.take());
    }

    static final class PriorityImpl {

        private Integer comp;

        public PriorityImpl(Integer comp) {
            this.comp = comp;
        }

        @Override
        public String toString() {
            return "PriorityImpl{" +
                    "comp=" + comp +
                    '}';
        }
    }
}
```

执行结果如下:

```
true
true
true
```

```
true
PriorityImpl{comp = 2}
PriorityImpl{comp = 15}
PriorityImpl{comp = 66}
PriorityImpl{comp = 78}
```

注意：上述代码运行后，最后一个 take()方法将导致当前执行线程阻塞等待。

7. poll(long timeout,TimeUnit unit)

检索并删除此队列的头部数据。接收 long 入参，作为最大等待时间，接收 TimeUnit 入参，作为时间单位。如果在最大等待给定时间段内检索成功，则返回删除的数据，否则返回 null，代码如下：

```java
//第7章/six/PriorityBlockingQueueTest.java
public class PriorityBlockingQueueTest {

    private static final PriorityBlockingQueue<PriorityImpl> QUEUE =
                            new PriorityBlockingQueue<>(16,
                            (o1, o2) -> o1.comp - o2.comp);

    public static void main(String[] args) throws Exception {
        System.out.println(QUEUE.offer(new PriorityImpl(15)));
        System.out.println(QUEUE.offer(new PriorityImpl(2)));
        System.out.println(QUEUE.add(new PriorityImpl(66)));
        System.out.println(QUEUE.add(new PriorityImpl(78)));

        System.out.println(QUEUE.take());
        System.out.println(QUEUE.take());
        System.out.println(QUEUE.take());
        System.out.println(QUEUE.take());
        System.out.println(QUEUE.poll(2, TimeUnit.SECONDS));
    }

    static final class PriorityImpl {

        private Integer comp;

        public PriorityImpl(Integer comp) {
            this.comp = comp;
        }

        @Override
        public String toString() {
            return "PriorityImpl{" +
                    "comp = " + comp +
                    '}';
        }
    }
}
```

执行结果如下：

```
true
true
true
true
PriorityImpl{comp = 2}
PriorityImpl{comp = 15}
PriorityImpl{comp = 66}
PriorityImpl{comp = 78}
Null
```

8. offer(E e, long timeout, TimeUnit unit)

添加数据，接收泛型入参，作为新加数据。由于此类是无界队列，所以此方法的底层会直接调用 offer(E e) 方法，如图 7-48 所示。

```
public boolean offer(E e, long timeout, TimeUnit unit) {
    return offer(e); // 永远不需要阻挡
}
```

图 7-48　offer 方法源代码

9. peek()

检索但不删除此队列的头部数据。如果成功，则返回检索的数据，如果此队列为空，则返回 null。

10. element()

检索但不删除此队列的头部数据。如果成功，则返回检索的数据，如果此队列为空，它将引发 NoSuchElementException 异常。

11. drainTo(Collection<? super E> c)

从此队列中删除所有可用元素，并将它们添加到给定集合中，然后返回添加的数量。接收 Collection 入参，作为给定集合。

12. remainingCapacity()

返回此队列在理想情况下还可以接受的元素数量，由于此类是无界队列，所以此方法固定返回 Integer.MAX_VALUE。

13. size()

返回此队列中已添加元素的数量，代码如下：

```java
//第 7 章/six/PriorityBlockingQueueTest.java
public class PriorityBlockingQueueTest {

    private static final PriorityBlockingQueue<PriorityImpl> QUEUE =
                                    new PriorityBlockingQueue<>(16,
                                    (o1, o2) -> o1.comp - o2.comp);

    public static void main(String[] args) throws Exception {
```

```
        System.out.println(QUEUE.offer(new PriorityImpl(15)));
        System.out.println(QUEUE.offer(new PriorityImpl(2)));
        System.out.println(QUEUE.add(new PriorityImpl(66)));
        System.out.println(QUEUE.add(new PriorityImpl(78)));

        System.out.println(QUEUE.size());
    }

    static final class PriorityImpl {

        private Integer comp;

        public PriorityImpl(Integer comp) {
            this.comp = comp;
        }

        @Override
        public String toString() {
            return "PriorityImpl{" +
                    "comp = " + comp +
                    '}';
        }
    }
}
```

执行结果如下：

```
true
true
true
true
4
```

小结

阻塞队列有多种实现形式，包括有界、无界、无容量、优先级、数组、链表、栈结构。读者需要理解各种形式之间存在的特性差异，根据业务需求选择合适的阻塞队列。

习题

1. 判断题

（1）无界队列任何生产数据的方法都没有阻塞效果。（　　）
（2）无界队列任何消费数据的方法都没有阻塞效果。（　　）
（3）有界队列任何生产数据的方法都没有阻塞效果。（　　）
（4）有界队列任何消费数据的方法都没有阻塞效果。（　　）

(5) 无容量队列任何生产数据的方法都有阻塞效果。（　　）
(6) 无容量队列任何消费数据的方法都有阻塞效果。（　　）

2. 选择题

(1) ArrayBlockingQueue 的（　　）方法可能有阻塞效果。（多选）
  A. offer(E e) B. poll() C. put(E e) D. take()

(2) LinkedTransferQueue 的（　　）方法可能有阻塞效果。（多选）
  A. transfer(E e) B. poll() C. put(E e) D. take()

(3) SynchronousQueue 的（　　）方法可能有阻塞效果。（多选）
  A. add(E e) B. poll() C. put(E e) D. take()

(4) 数据的有排序效果有（　　）类。（多选）
  A. DelayQueue   B. PriorityBlockingQueue
  C. SynchronousQueue   D. LinkedBlockingDeque

3. 填空题

(1) 根据业务要求补全代码，无容量队列要求正常生产并消费数据，程序正常结束，代码如下：

```java
//第7章/answer/NoCapacity.java
public class NoCapacity {

    private static final BlockingQueue<String> QUEUE = _____;

    public static void main(String[] args) throws Exception {
        BlockingRun blockingRun = new BlockingRun();
        new Thread(blockingRun).start();
        Thread.sleep(500);
        QUEUE.add("one");
    }

    static final class BlockingRun implements Runnable{

        @Override
        public void run() {
            try {
                System.out.println(QUEUE._____);
            } catch (InterruptedException e) {
                e.printStackTrace();
            }
        }
    }
}
```

(2) 查看执行结果，并补充代码，代码如下：

```java
//第7章/answer/PriorityAnswer.java
public class PriorityAnswer {
```

```
    private static final BlockingQueue< Integer > QUEUE = _____;

    public static void main(String[] args) throws Exception {
        QUEUE.add(50);
        QUEUE.add(100);
        QUEUE.add(2);
        QUEUE.add(456);
        QUEUE.add(6);
        System.out.println(QUEUE.take());
        System.out.println(QUEUE.take());
        System.out.println(QUEUE.take());
        System.out.println(QUEUE.take());
        System.out.println(QUEUE.take());
    }

}
```

执行结果如下：

```
2
6
50
100
456
```

第 8 章 线 程 池

129min

线程池的核心概念就是线程对象的复用，每次创建、销毁线程对象肯定都会有性能的损耗，在合理的范围内复用线程对象可以降低这种损耗，线程对象的复用也需要根据业务场景进行选择，并不是什么情况下都适合复用。

线程对象复用的简单演示，代码如下：

```java
//第8章/one/ThreadPoolLead.java
public class ThreadPoolLead {

    private static final BlockingQueue<QueueRun> BLOCKING_QUEUE =
                                    new ArrayBlockingQueue<>(64);

    private static final Random RANDOM = new Random();

    public static void main(String[] args) {
        LoopRun loopRun = new LoopRun();
        new Thread(loopRun,"A").start();
        new Thread(loopRun,"B").start();
        QueueRun one = new QueueRun("one");
        BLOCKING_QUEUE.add(one);
        BLOCKING_QUEUE.add(new QueueRun("two"));
        BLOCKING_QUEUE.add(new QueueRun("three"));
        BLOCKING_QUEUE.add(new QueueRun("four"));

        System.out.println(one.getResult());           //阻塞等待获得返回结果
        System.out.println("main-end");
    }

    static final class LoopRun implements Runnable{

        @Override
        public void run() {
            try {
                while (true){//死循环复用当前执行线程对象
                    QueueRun take = BLOCKING_QUEUE.take();
                    take.run();
                }
```

```java
        } catch (InterruptedException e) {
            e.printStackTrace();
        }
    }
}

static final class QueueRun implements Runnable{

    private String name;

    private Integer result;

    private Lock lock = new ReentrantLock();

    private Condition condition = lock.newCondition();

    public QueueRun(String name) {
        this.name = name;
    }

    @Override
    public void run() {
        lock.lock();
        try {
            Thread.sleep(RANDOM.nextInt(5000));       //随机睡眠一个时间段
            System.out.println(Thread.currentThread().getName()
                                               + ":" + this.name);
            result = RANDOM.nextInt(666666);
        } catch (InterruptedException e) {
            result = -1;
            e.printStackTrace();
        } finally {
            condition.signalAll();
            lock.unlock();
        }
    }

    public Integer getResult(){
        lock.lock();
        try {
          while (result == null){
              condition.await();               //阻塞等待任务完成
          }
          return result;
        } catch (InterruptedException e) {
            e.printStackTrace();
        }finally {
            lock.unlock();
        }
        return null;
```

```
            }
        }
    }
}
```

执行结果如下：

```
A:one
263446
main-end
B:two
A:three
B:four
```

注意：上述代码运行后，会复用 A、B 两个执行线程对象，所以 JVM 不会关闭。

8.1 ThreadPoolExecutor

提供了线程的复用、线程的管理、任务的管理、任务统计、任务结果获取等功能。

8.1.1 构造器

ThreadPoolExecutor 构造器见表 8-1。

表 8-1　ThreadPoolExecutor 构造器

构造器	描述
ThreadPoolExecutor（int corePoolSize，int maximumPoolSize，long keepAliveTime，TimeUnit unit，BlockingQueue < Runnable > workQueue）	构造新的对象，指定核心线程数量，指定最大线程数量，指定超时时间，指定时间单位，指定阻塞队列
ThreadPoolExecutor（int corePoolSize，int maximumPoolSize，long keepAliveTime，TimeUnit unit，BlockingQueue < Runnable > workQueue，RejectedExecutionHandler handler）	构造新的对象，指定核心线程数量，指定最大线程数量，指定超时时间，指定时间单位，指定阻塞队列，指定拒绝处理器
ThreadPoolExecutor（int corePoolSize，int maximumPoolSize，long keepAliveTime，TimeUnit unit，BlockingQueue < Runnable > workQueue，ThreadFactory threadFactory）	构造新的对象，指定核心线程数量，指定最大线程数量，指定超时时间，指定时间单位，指定阻塞队列，指定线程工厂
ThreadPoolExecutor（int corePoolSize，int maximumPoolSize，long keepAliveTime，TimeUnit unit，BlockingQueue < Runnable > workQueue，ThreadFactory threadFactory，RejectedExecutionHandler handler）	构造新的对象，指定核心线程数量，指定最大线程数量，指定超时时间，指定时间单位，指定阻塞队列，指定线程工厂，指定拒绝处理器

执行流程如图 8-1 所示。

```
public void execute(Runnable command) {
    if (command == null)
        throw new NullPointerException();
    int c = ctl.get();
    if (workerCountOf(c) < corePoolSize) {    ← 当前线程数量小于核心数量时
        if (addWorker(command, true))         ← 添加核心数量内Worker对象
            return;
        c = ctl.get();
    }
    if (isRunning(c) && workQueue.offer(command)) {  ← 核心线程数量已满，添加任务入队列
        int recheck = ctl.get();
        if (! isRunning(recheck) && remove(command))
            reject(command);
        else if (workerCountOf(recheck) == 0)
            addWorker(null, false);
    }
    else if (!addWorker(command, false))     ← 任务队列已满，添加最大数量内Worker对象
        reject(command);                      ← 所有容量已满，走拒绝处理器
}
```

图 8-1　execute 方法源代码

线程池内 workers 字段用于存储 Worker 对象集合，每个 Worker 对象内都有一个执行线程对象，新添加 Worker 对象时如果有直接任务，则优先执行此任务，后续会到阻塞队列中获取任务，如图 8-2 所示。

getTask() 方法会影响当前执行线程对象的清理机制，如图 8-3 所示。

注意：Worker 内部类是由非 static 修饰的，所以会有一个外部类对象引用，在 Worker 内部类对象方法里可以直接调用外部类对象的方法。

8.1.2　常用方法

1. allowCoreThreadTimeOut(boolean value)

设置策略，该策略用于控制在最大过期时间后没有任务到达时核心线程数量是否可以回收。接收 boolean 入参，作为策略控制，代码如下：

```java
//第8章/one/ThreadPoolExecutorTest.java
public class ThreadPoolExecutorTest {

    private static final ThreadPoolExecutor THREAD_POOL =
            new ThreadPoolExecutor(2,5,2, TimeUnit.SECONDS,
                    new ArrayBlockingQueue<>(5), new MyDiscardPolicy());

    public static void main(String[] args) throws InterruptedException {
```

```java
final void runWorker(Worker w) {
    Thread wt = Thread.currentThread();
    Runnable task = w.firstTask;        // 优先执行自身任务
    w.firstTask = null;
    w.unlock(); // 允许中断
    boolean completedAbruptly = true;
    try {
        while (task != null || (task = getTask()) != null) {   // 到阻塞队列获取任务
            w.lock();
            if ((runStateAtLeast(ctl.get(), STOP) ||
                 (Thread.interrupted() &&
                  runStateAtLeast(ctl.get(), STOP))) &&
                !wt.isInterrupted())
                wt.interrupt();
            try {
                beforeExecute(wt, task);    // 前置处理器
                try {
                    task.run();             // 执行任务本身
                    afterExecute(task, null);   // 后置处理器
                } catch (Throwable ex) {
                    afterExecute(task, ex);
                    throw ex;
                }
            } finally {
                task = null;                // 自身任务清理
                w.completedTasks++;         // 统计数量
                w.unlock();
            }
        }
        completedAbruptly = false;
    } finally {
        processWorkerExit(w, completedAbruptly);    // 清理流程
    }
}
```

图 8-2　runWorker 方法源代码

```java
private Runnable getTask() {
    boolean timedOut = false;
    for (;;) {
        int c = ctl.get();
        if (runStateAtLeast(c, SHUTDOWN)        // 状态检查
            && (runStateAtLeast(c, STOP) || workQueue.isEmpty())) {
            decrementWorkerCount();
            return null;    // 返回空，会导致当前执行线程正常结束生命周期
        }
        int wc = workerCountOf(c);      // 当前线程数量
        boolean timed = allowCoreThreadTimeOut || wc > corePoolSize;
        if ((wc > maximumPoolSize || (timed && timedOut))
            && (wc > 1 || workQueue.isEmpty())) {
            if (compareAndDecrementWorkerCount(c))
                return null;    // 返回空，会导致当前执行线程正常结束生命周期
            continue;
        }
        try {
            Runnable r = timed ?    // 阻塞等待最大时间段
                                    // 此处构造器将传入的超时时间转换为纳秒
                workQueue.poll(keepAliveTime, TimeUnit.NANOSECONDS) :
                workQueue.take();   // 阻塞等待
            if (r != null)
                return r;   // 如果得到了任务就返回此任务
            timedOut = true;
        } catch (InterruptedException retry) {
            timedOut = false;
        }
    }
}
```

图 8-3　getTask 方法源代码

```java
            for (int i = 0; i < 10; i++) {
                THREAD_POOL.execute(new Runnable() {
                    @Override
                    public void run() {
                        try {
                            //在实际业务中需要处理 interrupt
                            Thread.sleep(2000);
                            System.out.println(Thread.currentThread().getName());
                        } catch (InterruptedException e) {
                            e.printStackTrace();
                        }
                    }
                });
            }
            THREAD_POOL.allowCoreThreadTimeOut(true);
            System.out.println("main - end");
        }
        /*
         * 自定义拒绝处理器
         */
        public static class MyDiscardPolicy implements RejectedExecutionHandler {

            public MyDiscardPolicy() { }

            public void rejectedExecution(Runnable r, ThreadPoolExecutor e) {
                System.out.println("所有的容量已经满了,到了拒绝处理器了");
            }
        }
    }
```

执行结果如下:

```
main - end
pool - 1 - thread - 3
pool - 1 - thread - 2
pool - 1 - thread - 1
pool - 1 - thread - 5
pool - 1 - thread - 4
pool - 1 - thread - 3
pool - 1 - thread - 1
pool - 1 - thread - 2
pool - 1 - thread - 4
pool - 1 - thread - 5
```

注意:主执行线程执行完毕,由于设置了回收核心线程,所以线程池内所有线程都会超时回收,然后 JVM 关闭。

2. allowsCoreThreadTimeOut()

如果此线程池允许核心线程回收,则返回 true,否则返回 false。

3. awaitTermination(long timeout,TimeUnit unit)

阻塞等待，直到此线程池关闭、超时最大等待时间或者当前执行线程被中断。接收 long 入参，作为最大等待时间，接收 TimeUnit 入参，作为时间单位。如果在最大等待给定时间段内此线程池关闭，则返回值为 true，否则返回值为 false，代码如下：

```java
//第8章/one/ThreadPoolExecutorTest.java
public class ThreadPoolExecutorTest {

    private static final ThreadPoolExecutor THREAD_POOL =
            new ThreadPoolExecutor(2,5,2, TimeUnit.SECONDS,
                    new ArrayBlockingQueue<>(5));

    public static void main(String[] args) throws InterruptedException {
        for (int i = 0; i < 10; i++) {
            THREAD_POOL.execute(new Runnable() {
                @Override
                public void run() {
                    try {
                        //在实际业务中需要处理 interrupt
                        Thread.sleep(200);
                        System.out.println(Thread.currentThread().getName());
                    } catch (InterruptedException e) {
                        e.printStackTrace();
                    }
                }
            });
        }
        THREAD_POOL.shutdown();
        System.out.println(THREAD_POOL.awaitTermination(5
                , TimeUnit.SECONDS));
        System.out.println("main - end");
    }

}
```

执行结果如下：

```
pool - 1 - thread - 2
pool - 1 - thread - 4
pool - 1 - thread - 5
pool - 1 - thread - 3
pool - 1 - thread - 1
pool - 1 - thread - 3
pool - 1 - thread - 5
pool - 1 - thread - 2
pool - 1 - thread - 4
pool - 1 - thread - 1
true
main - end
```

4. execute(Runnable command)

将新的任务添加到此线程池内。接收 Runnable 入参，作为新的任务。

5. getActiveCount()

返回此线程池内正在执行任务的线程的大致数量，代码如下：

```java
//第 8 章/one/ThreadPoolExecutorTest.java
public class ThreadPoolExecutorTest {

    private static final ThreadPoolExecutor THREAD_POOL =
            new ThreadPoolExecutor(2,5,2, TimeUnit.SECONDS,
                    new ArrayBlockingQueue<>(5));

    public static void main(String[] args) throws InterruptedException {
        for (int i = 0; i < 10; i++) {
            THREAD_POOL.execute(new Runnable() {
                @Override
                public void run() {
                    try {
                        //在实际业务中需要处理 interrupt
                        Thread.sleep(2000);
                        System.out.println(Thread.currentThread().getName());
                    } catch (InterruptedException e) {
                        e.printStackTrace();
                    }
                }
            });
        }
        THREAD_POOL.allowCoreThreadTimeOut(true);
        System.out.println(THREAD_POOL.getActiveCount());
        System.out.println("main - end");
    }
}
```

执行结果如下：

```
5
main - end
pool - 1 - thread - 1
pool - 1 - thread - 3
pool - 1 - thread - 2
pool - 1 - thread - 4
pool - 1 - thread - 5
pool - 1 - thread - 3
pool - 1 - thread - 5
pool - 1 - thread - 4
pool - 1 - thread - 2
pool - 1 - thread - 1
```

6. getCompletedTaskCount()

返回已经完成的任务的大致总数量。

7. getCorePoolSize()

返回核心线程数量。

8. getKeepAliveTime(TimeUnit unit)

返回最大空闲超时时间，接收 TimeUnit 入参，作为时间单位。

9. getLargestPoolSize()

返回此线程池中已经创建过的最大线程数量，代码如下：

```java
//第 8 章/one/ThreadPoolExecutorTest.java
public class ThreadPoolExecutorTest {

    private static final ThreadPoolExecutor THREAD_POOL =
            new ThreadPoolExecutor(2,5,2, TimeUnit.SECONDS,
                    new ArrayBlockingQueue<>(1));

    public static void main(String[] args) throws InterruptedException {
        for (int i = 0; i < 4; i++) {
            THREAD_POOL.execute(new Runnable() {
                @Override
                public void run() {
                    try {
                        //在实际业务中需要处理 interrupt
                        Thread.sleep(2000);
                        System.out.println(Thread.currentThread().getName());
                    } catch (InterruptedException e) {
                        e.printStackTrace();
                    }
                }
            });
        }
        THREAD_POOL.allowCoreThreadTimeOut(true);
        System.out.println(THREAD_POOL.getLargestPoolSize());
        System.out.println("main - end");
    }

}
```

执行结果如下：

```
3
main - end
pool - 1 - thread - 1
pool - 1 - thread - 2
pool - 1 - thread - 3
pool - 1 - thread - 1
```

10. getMaximumPoolSize()

返回允许的最大线程数量。

11. getPoolSize()

返回池中的当前线程数量,代码如下:

```java
//第8章/one/ThreadPoolExecutorTest.java
public class ThreadPoolExecutorTest {

    private static final ThreadPoolExecutor THREAD_POOL =
            new ThreadPoolExecutor(2, 5, 2, TimeUnit.SECONDS,
                    new ArrayBlockingQueue<>(5));

    public static void main(String[] args) throws InterruptedException {
        for (int i = 0; i < 10; i++) {
            THREAD_POOL.execute(new Runnable() {
                @Override
                public void run() {
                    try {
                        //在实际业务中需要处理 interrupt
                        Thread.sleep(200);
                        System.out.println(Thread.currentThread().getName());
                    } catch (InterruptedException e) {
                        e.printStackTrace();
                    }
                }
            });
        }
        THREAD_POOL.allowCoreThreadTimeOut(true);
        Thread.sleep(3000);
        System.out.println(THREAD_POOL.getLargestPoolSize());
        System.out.println(THREAD_POOL.getCorePoolSize());
        System.out.println(THREAD_POOL.getPoolSize());
        System.out.println("main - end");
    }

}
```

执行结果如下:

```
pool - 1 - thread - 3
pool - 1 - thread - 2
pool - 1 - thread - 1
pool - 1 - thread - 5
pool - 1 - thread - 4
pool - 1 - thread - 3
pool - 1 - thread - 1
pool - 1 - thread - 2
pool - 1 - thread - 5
pool - 1 - thread - 4
5
2
0
main - end
```

12. getQueue()

返回此线程池内的阻塞任务队列。

13. getRejectedExecutionHandler()

返回此线程池内的拒绝处理器。

14. getTaskCount()

返回已执行的任务、没执行的任务、正在执行的任务,所有任务的总数量。

15. getThreadFactory()

返回此线程池内的线程工厂实现类。

16. isShutdown()

如果此线程池已执行关闭操作,则返回值为 true,否则返回值为 false。

17. isTerminated()

如果此线程池已终止完成,则返回值为 true,否则返回值为 false。

18. isTerminating()

如果此线程池正在终止中,则返回值为 true,否则返回值为 false。

19. prestartAllCoreThreads()

启动所有核心线程,使它们阻塞等待工作。

20. prestartCoreThread()

如果当前线程数量小于核心线程数量,则启动一个新的线程使它阻塞等待工作。

21. purge()

尝试从阻塞队列中删除所有已取消的 Future<V> 任务。

22. setCorePoolSize(int corePoolSize)

设置核心线程数量。接收 int 入参,作为核心数量。

23. setKeepAliveTime(long time, TimeUnit unit)

设置最大空闲超时时间,接收 long 入参,作为最大等待时间,接收 TimeUnit 入参,作为时间单位。

24. setMaximumPoolSize(int maximumPoolSize)

设置允许的最大线程数量。接收 int 入参,作为最大线程数量。

25. setRejectedExecutionHandler(RejectedExecutionHandler handler)

设置拒绝处理器,接收 RejectedExecutionHandler 入参,作为拒绝处理器。

26. setThreadFactory(ThreadFactory threadFactory)

设置线程工厂,接收 ThreadFactory 入参,作为线程工厂实现类。

27. shutdown()

启动有序关闭,在该关闭中执行已经提交的任务,但不接受任何新任务,代码如下:

```
//第 8 章/one/ThreadPoolExecutorTest.java
public class ThreadPoolExecutorTest {
```

```java
private static final ThreadPoolExecutor THREAD_POOL =
    new ThreadPoolExecutor(
            2,5,2, TimeUnit.SECONDS, new ArrayBlockingQueue<>(5)
            , new MyDiscardPolicy());

    public static void main(String[] args) throws InterruptedException {
        for (int i = 0; i < 10; i++) {
            THREAD_POOL.execute(new Runnable() {
                @Override
                public void run() {
                    try {
                        //在实际业务中需要处理 interrupt
                        Thread.sleep(2000);
                        System.out.println(Thread.currentThread().getName());
                    } catch (InterruptedException e) {
                        e.printStackTrace();
                    }
                }
            });
        }
        Thread.sleep(500);
        THREAD_POOL.shutdown();
        System.out.println("isShutdown:" + THREAD_POOL.isShutdown());
        System.out.println("isTerminating:" + THREAD_POOL.isTerminating());
        System.out.println("isTerminated:" + THREAD_POOL.isTerminated());
        System.out.println("main - end");
    }

    public static class MyDiscardPolicy implements RejectedExecutionHandler {

        public MyDiscardPolicy() { }

        public void rejectedExecution(Runnable r, ThreadPoolExecutor e) {
            System.out.println("所有的容量已经满了,到了拒绝处理器了");
        }
    }
}
```

执行结果如下：

```
isShutdown:true
isTerminating:true
isTerminated:false
main - end
pool - 1 - thread - 1
pool - 1 - thread - 2
pool - 1 - thread - 4
pool - 1 - thread - 3
pool - 1 - thread - 5
pool - 1 - thread - 2
```

```
pool-1-thread-1
pool-1-thread-3
pool-1-thread-5
pool-1-thread-4
```

28. shutdownNow()

尝试立即停止所有正在执行的任务,停止对阻塞队列任务的处理,删除阻塞队列中所有的任务,并返回删除的阻塞队列任务列表,代码如下:

```java
//第8章/one/ThreadPoolExecutorTest.java
public class ThreadPoolExecutorTest {

    private static final ThreadPoolExecutor THREAD_POOL =
        new ThreadPoolExecutor(
                2,5,2, TimeUnit.SECONDS, new ArrayBlockingQueue<>(5),
                new MyDiscardPolicy());

    public static void main(String[] args) throws InterruptedException {
        for (int i = 0; i < 10; i++) {
            THREAD_POOL.execute(new Runnable() {
                @Override
                public void run() {
                    try {
                        //在实际业务中需要处理 interrupt
                        Thread.sleep(2000);
                        System.out.println(Thread.currentThread().getName());
                    } catch (InterruptedException e) {
                        System.out.println(Thread.currentThread().getName()
                                        + ":InterruptedException");
                    }
                }
            });
        }
        Thread.sleep(500);
        System.out.println(THREAD_POOL.shutdownNow());
        System.out.println("isShutdown:" + THREAD_POOL.isShutdown());
        System.out.println("isTerminating:" + THREAD_POOL.isTerminating());
        Thread.sleep(6000);
        System.out.println("isTerminated:" + THREAD_POOL.isTerminated());
        System.out.println("main-end");
    }

    public static class MyDiscardPolicy implements RejectedExecutionHandler {

        public MyDiscardPolicy() { }

        public void rejectedExecution(Runnable r, ThreadPoolExecutor e) {
            System.out.println("所有的容量已经满了,到了拒绝处理器了");
        }
    }
}
```

执行结果如下:

```
[cn.kungreat.book.eight.one.ThreadPoolExecutorTest $ 1@5fd0d5ae,
cn.kungreat.book.eight.one.ThreadPoolExecutorTest $ 1@2d98a335,
cn.kungreat.book.eight.one.ThreadPoolExecutorTest $ 1@16b98e56,
cn.kungreat.book.eight.one.ThreadPoolExecutorTest $ 1@7ef20235,
cn.kungreat.book.eight.one.ThreadPoolExecutorTest $ 1@27d6c5e0]
pool-1-thread-4:InterruptedException
pool-1-thread-2:InterruptedException
pool-1-thread-1:InterruptedException
pool-1-thread-5:InterruptedException
pool-1-thread-3:InterruptedException
isShutdown:true
isTerminating:false
isTerminated:true
main-end
```

8.2 FutureTask

此类实现了 RunnableFuture＜V＞接口,并提供了 Runnable、Future＜V＞接口的基本实现,具有启动任务、取消任务、查询任务是否完成、检索任务结果的功能。

8.2.1 构造器

FutureTask 构造器见表 8-2。

表 8-2 FutureTask 构造器

构造器	描述
FutureTask(Runnable runnable,V result)	构造新的对象,指定 Runnable 实现类,指定返回结果
FutureTask(Callable＜V＞ callable)	构造新的对象,指定 Callable 实现类

核心字段说明,如图 8-4 所示。

在使用 FutureTask(Runnable runnable,V result)构造器时,会通过 Executors.callable (runnable,result)方法返回包装的 Callable 实现类,如图 8-5 所示。

包装的 Callable 实现类源代码,如图 8-6 所示。

8.2.2 常用方法

1. run()

执行此任务,此方法是实现 Runnable 接口的方法,如图 8-7 所示。

2. get()

阻塞等待任务执行完成,然后返回结果,代码如下:

```java
private volatile int state;//状态
private static final int NEW          = 0;//新创建任务
private static final int COMPLETING   = 1;//任务中转状态
private static final int NORMAL       = 2;//任务正常结束
private static final int EXCEPTIONAL  = 3;//任务异常结束
private static final int CANCELLED    = 4;//任务取消
private static final int INTERRUPTING = 5;//任务取消并处于中断中
private static final int INTERRUPTED  = 6;//任务取消并中断完成

private Callable<V> callable;
/** 返回的结果 */
private Object outcome;
/** 当前任务的执行线程 */
private volatile Thread runner;
/** 阻塞等待任务结果的链表 */
private volatile WaitNode waiters;
```

图 8-4 FutureTask 核心字段

```java
public FutureTask(Runnable runnable, V result) {
    this.callable = Executors.callable(runnable, result);
    this.state = NEW;      //初始状态
}
```

图 8-5 构造器源代码

```java
private static final class RunnableAdapter<T> implements Callable<T> {
    private final Runnable task;
    private final T result;
    RunnableAdapter(Runnable task, T result) {
        this.task = task;
        this.result = result;
    }
    public T call() {
        task.run();
        return result;
    }
    public String toString() {
        return super.toString() + "[Wrapped task = " + task + "]";
    }
}
```

图 8-6 RunnableAdapter 类源代码

```java
public void run() {
    if (state != NEW ||
        !RUNNER.compareAndSet(this, null, Thread.currentThread()))   // 设置当前任务的执行线程
        return;
    try {
        Callable<V> c = callable;
        if (c != null && state == NEW) {
            V result;
            boolean ran;
            try {
                result = c.call();// 回调call()方法
                ran = true;
            } catch (Throwable ex) {
                result = null;// 当前任务的执行线程置空
                ran = false;
                setException(ex);//任务异常执行完时
            }
            if (ran)
                set(result);//任务正常执行完时
        }
    } finally {
        // 当前任务的执行线程置空
        runner = null;
        int s = state;
        if (s >= INTERRUPTING)
            handlePossibleCancellationInterrupt(s);
    }
}
```

图 8-7 run()方法源代码

```java
//第 8 章/two/FutureTaskTest.java
public class FutureTaskTest {

    private static final FutureTask<Integer> FUTURE_TASK =
            new FutureTask<>(new RunnableImpl(),100);

    public static void main(String[] args) throws Exception {
        new Thread(FUTURE_TASK,"A").start();
        new Thread(() -> {
            try {
                System.out.println(FUTURE_TASK.get());
                                        //阻塞等待任务执行完成,然后返回结果
            } catch (Exception e) {
                e.printStackTrace();
            }
        },"B").start();
        System.out.println("main-end");
    }

    static class RunnableImpl implements Runnable{

        @Override
        public void run() {
            try {
```

```
                    Thread.sleep(5000);
                    System.out.println("RunnableImpl:"
                            + Thread.currentThread().getName());
            } catch (InterruptedException e) {
                e.printStackTrace();
            }
        }
    }
}
```

执行结果如下：

```
main - end
RunnableImpl:A
100
```

get()方法只有在状态为 NORMAL 时才能正常返回结果，否则将抛出异常，如图 8-8 所示。

```
public V get() throws InterruptedException, ExecutionException {
    int s = state;
    if (s <= COMPLETING)
        s = awaitDone(false, 0L);  ← 任务没有完成时阻塞等待
    return report(s);
}
@SuppressWarnings("unchecked")
private V report(int s) throws ExecutionException {
    Object x = outcome;
    if (s == NORMAL)
        return (V)x;  ← 此状态时才正常返回结果
    if (s >= CANCELLED)
        throw new CancellationException();
    throw new ExecutionException((Throwable)x);
}
```

图 8-8　get()方法源代码(1)

3. get(long timeout,TimeUnit unit)

最大时间阻塞等待任务执行完成，然后返回结果。接收 long 入参，作为最大等待时间，接收 TimeUnit 入参，作为时间单位。在任务超过最大等待时间后还没有完成的情况下将抛出 TimeoutException 异常，代码如下：

```
//第 8 章/two/FutureTaskTest.java
public class FutureTaskTest {

    private static final FutureTask<Integer> FUTURE_TASK =
            new FutureTask<>(new RunnableImpl(),100);
```

```java
public static void main(String[] args) throws Exception {
    new Thread(FUTURE_TASK,"A").start();
    new Thread(() -> {
        try {
            System.out.println(FUTURE_TASK.get(2, TimeUnit.SECONDS));
        } catch (Exception e) {
            e.printStackTrace();
        }
    },"B").start();
    System.out.println("main-end");
}

static class RunnableImpl implements Runnable{

    @Override
    public void run() {
        try {
            Thread.sleep(5000);
            System.out.println("RunnableImpl:"
                    + Thread.currentThread().getName());
        } catch (InterruptedException e) {
            e.printStackTrace();
        }
    }
}
```

执行结果如下：

```
main-end
java.util.concurrent.TimeoutException
    at java.base/java.util.concurrent.FutureTask.get
(FutureTask.java:204)
    at cn.kungreat.book.eight.two.FutureTaskTest.lambda$main$0
(FutureTaskTest.java:16)
    at java.base/java.lang.Thread.run(Thread.java:833)
RunnableImpl:A
```

源代码如图8-9所示。

```java
    public V get(long timeout, @NotNull TimeUnit unit)
        throws InterruptedException, ExecutionException, TimeoutException {
        if (unit == null)
            throw new NullPointerException();
        int s = state;
        if (s <= COMPLETING &&
            (s = awaitDone(timed: true, unit.toNanos(timeout))) <= COMPLETING)
            throw new TimeoutException();   ←── 超时没有完成任务时
        return report(s);
    }
```

图8-9 get()方法源代码(2)

4. cancel(boolean mayInterruptIfRunning)

取消此任务,只有状态为 NEW 的任务才能取消,成功后返回值为 true,失败后返回值为 false。接收 boolean 入参,作为是否中断此任务执行线程的条件,代码如下:

```java
public class FutureTaskTest {

    private static final FutureTask<Integer> FUTURE_TASK =
                new FutureTask<>(new RunnableImpl(),100);

    public static void main(String[] args) throws Exception {
        new Thread(FUTURE_TASK,"A").start();
        new Thread(() -> {
            try {
                System.out.println(FUTURE_TASK.get());
            } catch (Exception e) {
                e.printStackTrace();
            }
        },"B").start();
        Thread.sleep(500);
        FUTURE_TASK.cancel(false);
        System.out.println("main-end");
    }

    static class RunnableImpl implements Runnable{

        @Override
        public void run() {
            try {
                Thread.sleep(5000);
                System.out.println("RunnableImpl:"
                            + Thread.currentThread().getName());
            } catch (InterruptedException e) {
                e.printStackTrace();
            }
        }
    }
}
```

执行结果如下:

```
main-end
java.util.concurrent.CancellationException
    at java.base/java.util.concurrent.FutureTask.report
(FutureTask.java:121)
    at java.base/java.util.concurrent.FutureTask.get
(FutureTask.java:191)
    at cn.kungreat.book.eight.two.FutureTaskTest.lambda$main$0
(FutureTaskTest.java:16)
    at java.base/java.lang.Thread.run(Thread.java:833)
RunnableImpl:A
```

> 注意：修改以上代码 FUTURE_TASK.cancel(true)，运行主方法并观察执行结果。

cancel(boolean mayInterruptIfRunning)方法的源代码如图 8-10 所示。

```
public boolean cancel(boolean mayInterruptIfRunning) {
    if (!(state == NEW && STATE.compareAndSet      ← 原子修改状态
          (this, NEW, mayInterruptIfRunning ? INTERRUPTING : CANCELLED)))
        return false;
    try {
        if (mayInterruptIfRunning) {
            try {
                Thread t = runner;
                if (t != null)
                    t.interrupt();                  ← 中断此任务的执行线程
            } finally { // final state
                STATE.setRelease(this, INTERRUPTED); ← 原子修改状态
            }
        }
    } finally {
        finishCompletion();                          ← 任务完成后唤醒阻塞等待的线程
    }
    return true;
}
```

图 8-10　cancel 方法源代码

5. isCancelled()

如果此任务被取消，则返回值为 true，否则返回值为 false。

6. isDone()

如果此任务已完成，则返回值为 true，否则返回值为 false。已完成包括正常完成、异常完成、任务取消 3 种状态。源代码如图 8-11 所示。

```
public boolean isDone() {
    return state != NEW;
}
```

图 8-11　isDone()源代码

8.3　AbstractExecutorService

此类是一个抽象类，线程池继承了此类，此类提供了 RunnableFuture 任务的实现，底层实现基于 FutureTask 类。

8.3.1　构造器

AbstractExecutorService 构造器见表 8-3。

表 8-3　AbstractExecutorService 构造器

构 造 器	说　　明
AbstractExecutorService()	构造新的对象，默认为无参构造器

8.3.2 常用方法

1. submit(Runnable task)

将任务添加到线程池,并返回表示该任务的 Future<?> 实现类。接收 Runnable 入参,作为新的任务,代码如下:

```java
//第8章/three/AbstractExecutorServiceTest.java
public class AbstractExecutorServiceTest {
    private static final AbstractExecutorService THREAD_POOL =
        new ThreadPoolExecutor(
                2,5,2, TimeUnit.SECONDS, new ArrayBlockingQueue<>(5));

    public static void main(String[] args) throws Exception {
        Future<?> submit = THREAD_POOL.submit(new RunnableImpl());
        System.out.println("main:" + submit.get());
    }

    static class RunnableImpl implements Runnable{

        @Override
        public void run() {
            try {
                Thread.sleep(5000);
                System.out.println("RunnableImpl:"
                        + Thread.currentThread().getName());
            } catch (InterruptedException e) {
                e.printStackTrace();
            }
        }
    }
}
```

执行结果如下:

```
RunnableImpl:pool-1-thread-1
main:null
```

观察 submit(Runnable task) 方法的源代码,会发现返回的 RunnableFuture 对象的底层使用了 FutureTask 实现类,如图 8-12 所示。

```java
public Future<?> submit( @NotNull Runnable task) {
    if (task == null) throw new NullPointerException();
    RunnableFuture<Void> ftask = newTaskFor(task, value: null);
    execute(ftask);    ← 将任务添加到线程池
    return ftask;      ← 返回RunnableFuture实现类
}
```

图 8-12 submit 源代码

2. submit(Runnable task,T result)

将任务添加到线程池,并返回表示该任务的 Future<T>实现类。接收 Runnable 入参,作为新的任务,接收泛型入参,作为检索 Future<T>的返回结果,代码如下:

```java
//第8章/three/AbstractExecutorServiceTest.java
public class AbstractExecutorServiceTest {
    private static final AbstractExecutorService THREAD_POOL =
    new ThreadPoolExecutor(
            2,5,2, TimeUnit.SECONDS, new ArrayBlockingQueue<>(5));

    public static void main(String[] args) throws Exception {
        Future<Integer> submit = THREAD_POOL.submit(new RunnableImpl(),100);
        System.out.println("main:" + submit.get());
    }

    static class RunnableImpl implements Runnable{

        @Override
        public void run() {
            try {
                Thread.sleep(5000);
                System.out.println("RunnableImpl:"
                        + Thread.currentThread().getName());
            } catch (InterruptedException e) {
                e.printStackTrace();
            }
        }
    }
}
```

执行结果如下:

```
RunnableImpl:pool-1-thread-1
main:100
```

3. submit(Callable<T> task)

将任务添加到线程池,并返回表示该任务的 Future<T>实现类。接收 Callable<T> 入参,作为新的任务,代码如下:

```java
//第8章/three/AbstractExecutorServiceTest.java
public class AbstractExecutorServiceTest {
    private static final Random RANDOM = new Random();
    private static final ThreadPoolExecutor THREAD_POOL =
    new ThreadPoolExecutor(
            2,5,2, TimeUnit.SECONDS, new ArrayBlockingQueue<>(5));

    public static void main(String[] args) throws Exception {
        Future<Integer> submit = THREAD_POOL.submit(
```

```java
                            new RunnableAdapter(new RunnableImpl()));
            System.out.println("main:" + submit.get());
        }

        static class RunnableImpl implements Runnable{
            @Override
            public void run() {
                try {
                    Thread.sleep(5000);
                    System.out.println("RunnableImpl:"
                                    + Thread.currentThread().getName());
                } catch (InterruptedException e) {
                    e.printStackTrace();
                }
            }
        }

        private static final class RunnableAdapter implements Callable< Integer >{
            private final Runnable task;
            RunnableAdapter(Runnable task) {
                this.task = task;
            }
            public Integer call() {
                task.run();
                return RANDOM.nextInt();
            }

            public String toString() {
                return super.toString() + "[Wrapped task = " + task + "]";
            }
        }
    }
```

执行结果如下：

```
RunnableImpl:pool-1-thread-1
main:891930084
```

4. invokeAny(Collection<? extends Callable< T >> tasks)

将集合任务添加到线程池,并返回第 1 个正常完成的 Future < T >任务结果。接收 Collection 入参,作为集合任务,代码如下：

```java
//第 8 章/three/AbstractExecutorServiceTestTwo.java
public class AbstractExecutorServiceTestTwo {
```

```java
    private static final ThreadPoolExecutor THREAD_POOL =
            new ThreadPoolExecutor(
            2,5,2, TimeUnit.SECONDS,
                    new ArrayBlockingQueue<>(5));

    public static void main(String[] args) throws Exception {

        Callable<String> callableOne = new Callable<String>() {
            @Override
            public String call() throws Exception {
                Thread.sleep(500);
                int ax = 1 / 0;
                return "one";
            }
        };

        Callable<String> callableTwo = new Callable<String>() {
            @Override
            public String call() throws Exception {
                Thread.sleep(2000);
                int ax = 1 / 0;
                return "two";
            }
        };

        Callable<String> callableThree = new Callable<String>() {
            @Override
            public String call() throws Exception {
                Thread.sleep(3000);
                return "three";
            }
        };

        List<Callable<String>> callableList =
                List.of(callableOne,callableTwo,callableThree);
        System.out.println(THREAD_POOL.invokeAny(callableList));

    }
}
```

执行结果如下：

```
three
```

5. invokeAll(Collection<? extends Callable<T>> tasks)

将集合任务添加到线程池，并等待所有任务完成，然后返回 List<Future<T>>。接收 Collection 入参，作为集合任务，代码如下：

```java
//第8章/three/AbstractExecutorServiceTestTwo.java
public class AbstractExecutorServiceTestTwo {
    private static final ThreadPoolExecutor THREAD_POOL =
    new ThreadPoolExecutor(
            2,5,2, TimeUnit.SECONDS, new ArrayBlockingQueue<>(5));

    public static void main(String[] args) throws Exception {
        Callable<String> callableOne = new Callable<String>() {
            @Override
            public String call() throws Exception {
                Thread.sleep(2000);
                int ax = 1 / 0;
                return "one";
            }
        };

        Callable<String> callableTwo = new Callable<String>() {
            @Override
            public String call() throws Exception {
                Thread.sleep(5000);
                return "two";
            }
        };

        Callable<String> callableThree = new Callable<String>() {
            @Override
            public String call() throws Exception {
                Thread.sleep(1000);
                return "three";
            }
        };

        List<Callable<String>> callableList =
                List.of(callableOne,callableTwo,callableThree);
        List<Future<String>> futures = THREAD_POOL.invokeAll(callableList);

        for (int i = 0; i < futures.size(); i++) {
            try {
                System.out.println(futures.get(i).get());
            }catch (Exception e){
                e.printStackTrace();
            }
        }
    }
}
```

执行结果如下：

```
java.util.concurrent.ExecutionException: java.lang.ArithmeticException: / by zero
    at
java.base/java.util.concurrent.FutureTask.report(FutureTask.java:122)
```

```
        at
java.base/java.util.concurrent.FutureTask.get(FutureTask.java:191)
        at
cn.kungreat.book.eight.three.AbstractExecutorServiceTestTwo.main
(AbstractExecutorServiceTestTwo.java:41)
Caused by: java.lang.ArithmeticException: / by zero
        at
cn.kungreat.book.eight.three.AbstractExecutorServiceTestTwo$1.call
(AbstractExecutorServiceTestTwo.java:15)
        at
cn.kungreat.book.eight.three.AbstractExecutorServiceTestTwo$1.call
(AbstractExecutorServiceTestTwo.java:11)
        at
java.base/java.util.concurrent.FutureTask.run(FutureTask.java:264)
        at
java.base/java.util.concurrent.ThreadPoolExecutor.runWorker(ThreadPoolExecutor.java:1136)
        at
java.base/java.util.concurrent.ThreadPoolExecutor$Worker.run(ThreadPoolExecutor.java:635)
        at java.base/java.lang.Thread.run(Thread.java:833)
two
three
```

8.4 ScheduledThreadPoolExecutor

此类继承了 ThreadPoolExecutor 类,并提供了延迟执行任务和周期性执行任务功能。

8.4.1 构造器

ScheduledThreadPoolExecutor 构造器见表 8-4。

表 8-4 ScheduledThreadPoolExecutor 构造器

构 造 器	说 明
ScheduledThreadPoolExecutor(int corePoolSize)	构造新的对象,指定核心线程数量,指定最大线程数量,指定超时时间,指定时间单位,指定阻塞队列
ScheduledThreadPoolExecutor(int corePoolSize, RejectedExecutionHandler handler)	构造新的对象,指定核心线程数量,指定拒绝处理器
ScheduledThreadPoolExecutor(int corePoolSize, ThreadFactory threadFactory)	构造新的对象,指定核心线程数量,指定线程工厂
ScheduledThreadPoolExecutor(int corePoolSize, ThreadFactory threadFactory, RejectedExecutionHandler handler)	构造新的对象,指定核心线程数量,指定线程工厂,指定拒绝处理器

8.4.2 常用方法

1. execute(Runnable command)

以零延迟将任务添加到阻塞队列中。接收 Runnable 入参,作为新的任务,代码如下:

```java
//第8章/four/ScheduledThreadTest.java
public class ScheduledThreadTest {

    static final ScheduledThreadPoolExecutor SCHEDULED_POOL =
                                    new ScheduledThreadPoolExecutor(2);

    public static void main(String[] args) {
        SCHEDULED_POOL.execute(new RunnableImpl());
    }

    static class RunnableImpl implements Runnable{

        @Override
        public void run() {
            try {
                Thread.sleep(5000);
                System.out.println(Thread.currentThread().getName());
            } catch (Exception e) {
                e.printStackTrace();
            }
        }
    }
}
```

执行结果如下:

```
pool-1-thread-1
```

2. getContinueExistingPeriodicTasksAfterShutdownPolicy()

获得有关是否执行现有定期任务的策略,返回 boolean 值。

3. getExecuteExistingDelayedTasksAfterShutdownPolicy()

获得有关是否执行现有延迟任务的策略,返回 boolean 值。

4. getQueue()

返回线程池所使用的阻塞队列,代码如下:

```java
//第8章/four/ScheduledThreadTest.java
public class ScheduledThreadTest {

    static final ScheduledThreadPoolExecutor SCHEDULED_POOL =
                    new ScheduledThreadPoolExecutor(2);

    public static void main(String[] args) {
        System.out.println(SCHEDULED_POOL.getQueue().getClass());
    }

}
```

执行结果如下:

```
class java.util.concurrent.ScheduledThreadPoolExecutor $ DelayedWorkQueue
```

注意：阻塞队列使用的是 ScheduledThreadPoolExecutor 类里面的 DelayedWorkQueue 静态内部类。

5. getRemoveOnCancelPolicy()
获得有关在取消时是否应立即从工作队列中删除已取消任务的策略，返回 boolean 值。

6. schedule(Runnable command, long delay, TimeUnit unit)
提交在给定延迟后执行的一次性任务。接收 Runnable 入参，作为新的任务，接收 long 入参，作为延迟时间，接收 TimeUnit 入参，作为时间单位。返回 ScheduledFuture<?>实现类，代码如下：

```java
//第8章/four/ScheduledThreadTest.java
public class ScheduledThreadTest {

    static final ScheduledThreadPoolExecutor SCHEDULED_POOL =
                                    new ScheduledThreadPoolExecutor(2);

    public static void main(String[] args) {
        SCHEDULED_POOL.schedule(new RunnableImpl(), 5, TimeUnit.SECONDS);
                                                //延迟5s后执行任务
    }

    static class RunnableImpl implements Runnable{
        @Override
        public void run() {
            try {
                System.out.println(Thread.currentThread().getName());
            } catch (Exception e) {
                e.printStackTrace();
            }
        }
    }
}
```

执行结果如下：

```
pool-1-thread-1
```

7. schedule(Callable< V > callable, long delay, TimeUnit unit)
提交在给定延迟后执行的一次性任务。接收 Callable < V >入参，作为新的任务，接收 long 入参，作为延迟时间，接收 TimeUnit 入参，作为时间单位。返回 ScheduledFuture< V >实现类，代码如下：

```java
//第8章/four/ScheduledThreadTest.java
public class ScheduledThreadTest {

    static final ScheduledThreadPoolExecutor SCHEDULED_POOL =
                                  new ScheduledThreadPoolExecutor(2);

    public static void main(String[] args) {
        SCHEDULED_POOL.schedule(new Callable<Object>() {
            @Override
            public Object call() throws Exception {
                System.out.println(Thread.currentThread().getName());
                return null;
            }
        }, 5, TimeUnit.SECONDS);
    }
}
```

执行结果如下:

```
pool-1-thread-1
```

8. scheduleAtFixedRate(Runnable command,long initialDelay,long period,TimeUnit unit)

提交一个定时任务,该任务在给定的初始延迟时间后首次执行,随后在到达间隔时间后重复执行此任务。接收 Runnable 入参,作为新的任务,接收 long 入参,作为延迟时间,接收 long 入参,作为间隔时间,接收 TimeUnit 入参,作为时间单位,代码如下:

```java
//第8章/four/ScheduledThreadTest.java
public class ScheduledThreadTest {

    static final ScheduledThreadPoolExecutor SCHEDULED_POOL =
                                  new ScheduledThreadPoolExecutor(2);

    public static void main(String[] args) throws Exception {
        SCHEDULED_POOL.scheduleAtFixedRate(new RunnableImpl()
                                      , 5,2, TimeUnit.SECONDS);
        Thread.sleep(20000);
        SCHEDULED_POOL.shutdown(); //有序关闭线程池
    }

    static class RunnableImpl implements Runnable{

        @Override
        public void run() {
            try {
                System.out.println(System.currentTimeMillis());
            } catch (Exception e) {
                e.printStackTrace();
            }
        }
    }
}
```

执行结果如下：

```
1674635103235
1674635105233
1674635107234
1674635109234
1674635111233
1674635113234
1674635115233
1674635117234
```

9. scheduleWithFixedDelay(Runnable command,long initialDelay,long delay,TimeUnit unit)

提交一个定时任务，该任务在给定的初始延迟时间后首次执行，随后在到达间隔时间后重复执行此任务。接收 Runnable 入参，作为新的任务，接收 long 入参，作为延迟时间，接收 long 入参，作为间隔时间，接收 TimeUnit 入参，作为时间单位，代码如下：

```java
//第8章/four/ScheduledThreadTest.java
public class ScheduledThreadTest {

    static final ScheduledThreadPoolExecutor SCHEDULED_POOL =
                                    new ScheduledThreadPoolExecutor(2);

    public static void main(String[] args) throws Exception {
        SCHEDULED_POOL.scheduleWithFixedDelay(new RunnableImpl(),
                                    5,2,TimeUnit.SECONDS);
        Thread.sleep(20000);
        SCHEDULED_POOL.shutdown(); //有序关闭线程池
    }

    static class RunnableImpl implements Runnable{

        @Override
        public void run() {
            try {
                System.out.println(System.currentTimeMillis());
            } catch (Exception e) {
                e.printStackTrace();
            }
        }
    }
}
```

执行结果如下：

```
1674635287070
1674635289071
1674635291072
1674635293074
1674635295075
```

```
1674635297076
1674635299078
1674635301079
```

10. setContinueExistingPeriodicTasksAfterShutdownPolicy(boolean value)

设置有关是否执行现有定期任务的策略。

11. setExecuteExistingDelayedTasksAfterShutdownPolicy(boolean value)

设置有关是否执行现有延迟任务的策略。

12. setRemoveOnCancelPolicy(boolean value)

设置有关在取消时是否应立即从工作队列中删除已取消任务的策略。

13. shutdown()

启动有序关闭,在该关闭中执行已经提交的任务,但不接受任何新任务。

14. shutdownNow()

尝试立即停止所有正在执行的任务,停止对阻塞队列任务的处理,删除阻塞队列中的所有任务,并返回删除的阻塞队列任务列表。

15. submit(Runnable task)

以零延迟将任务添加到阻塞队列中,并返回表示该任务的 Future<?>实现类。接收 Runnable 入参,作为新的任务。

16. submit(Runnable task,T result)

以零延迟将任务添加到阻塞队列中,并返回表示该任务的 Future<T>实现类。接收 Runnable 入参,作为新的任务,接收泛型入参,作为检索 Future<T>的返回结果。

17. submit(Callable<T> task)

以零延迟将任务添加到阻塞队列中,并返回表示该任务的 Future<T>实现类。接收 Callable<T>入参,作为新的任务。

小结

了解线程池的核心入参,在使用时根据业务场景选择合适的入参才能更好地发挥线程池的作用。了解 Future<V>的核心知识点,在主流框架中会大量用这种机制来完成任务的通知或等待。了解两种定时器的间隔时间算法,选择合适的定时方案。

习题

1. 判断题

(1) 线程池的核心线程数量和最大线程数量可以一样。( )

(2) 线程池的拒绝处理器可以自定义实现。( )

(3) 线程池的线程工厂类可以自定义实现。( )

(4) FutureTask<V>实现了 Future<V>接口。（   ）
(5) 线程池的阻塞队列可以自定义实现。（   ）
(6) 定时器的定时任务间隔算法默认有两种模式。（   ）

2．选择题

(1) ScheduledThreadPoolExecutor 的(   )方法有定时执行效果。（多选）

    A．submit(Runnable task)

    B．execute(Runnable command)

    C．scheduleAtFixedRate（Runnable command，long initialDelay，long period，TimeUnit unit）

    D．scheduleWithFixedDelay（Runnable command，long initialDelay，long delay，TimeUnit unit）

(2) ThreadPoolExecutor 的(   )方法能返回 Future 任务实现。（多选）

    A．execute(Runnable command)   B．prestartCoreThread()

    C．submit(Runnable task)   D．submit(Callable<T> task)

(3) Future 的(   )方法有等待此任务完成并获得任务结果的功能。（单选）

    A．get()   B．get(long timeout，TimeUnit unit)

    C．isCancelled()   D．isDone()

3．填空题

(1) 根据业务要求补全代码，定时任务首次延迟时间为 10s，间隔执行时间为 3s，代码如下：

```java
//第 8 章/answer/ScheduledThread.java
public class ScheduledThread {

    static final ScheduledThreadPoolExecutor SCHEDULED_POOL =
                                    new ScheduledThreadPoolExecutor(2);

    public static void main(String[] args) {
        SCHEDULED_POOL.scheduleAtFixedRate(new RunnableImpl(),_____);
    }

    static class RunnableImpl implements Runnable{
        @Override
        public void run() {
            try {
                System.out.println(System.currentTimeMillis());
            } catch (Exception e) {
                e.printStackTrace();
            }
        }
    }
}
```

(2) 根据业务要求补全代码，线程池使用有延迟效果和排序效果的阻塞队列，代码如下：

```java
//第8章/answer/DelayThread.java
public class DelayThread {

    private static final ThreadPoolExecutor THREAD_POOL =
        new ThreadPoolExecutor(1,1,20, TimeUnit.SECONDS,_____);

    public static void main(String[] args) {
        THREAD_POOL._____;
        long currentTimeMillis = System.currentTimeMillis();
        THREAD_POOL.execute(new DelayedImpl(20,currentTimeMillis + 3000));
        THREAD_POOL.execute(new DelayedImpl(5,currentTimeMillis + 5000));
        THREAD_POOL.execute(new DelayedImpl(3,currentTimeMillis + 10000));
        THREAD_POOL.execute(new DelayedImpl(80,currentTimeMillis + 2000));
        THREAD_POOL.execute(new DelayedImpl(160,currentTimeMillis + 2000));
    }

    static final class DelayedImpl implements Runnable, Delayed {
        private Integer comp;
        private Long delayTime;

        public DelayedImpl(Integer comp, Long delayTime) {
            this.comp = comp;
            this.delayTime = delayTime;
        }

        @Override
        public long getDelay(TimeUnit unit) {
            return delayTime - System.currentTimeMillis();
        }

        @Override
        public int compareTo(Delayed o) {
            DelayedImpl delayed = (DelayedImpl) o;
            return comp - delayed.comp;
        }

        @Override
        public void run() {
            try {
                System.out.println(this);
            } catch (Exception e) {
                e.printStackTrace();
            }
        }

        @Override
        public String toString() {
            return "DelayedImpl{" +
                "comp = " + comp +
```

```
                    }';
        }

    }
}
```

执行结果如下：

```
DelayedImpl{comp = 3}
DelayedImpl{comp = 5}
DelayedImpl{comp = 20}
DelayedImpl{comp = 80}
DelayedImpl{comp = 160}
```

第 9 章 线程同步器

线程同步器可以协调多个线程之间的工作,官方提供的工具类可以很方便地实现协调功能。

9.1 CountDownLatch

一种同步辅助工具,它接受一个初始计数器,并允许一个或多个执行线程等待,直到此计数器为 0 时唤醒所有等待的线程。

9.1.1 构造器

CountDownLatch 构造器见表 9-1。

表 9-1 CountDownLatch 构造器

构 造 器	说 明
CountDownLatch(int count)	构造新的对象,指定初始计数器

9.1.2 常用方法

1. await()

使当前执行线程阻塞等待,直到计数器为 0 或者此线程中断,代码如下:

```
//第 9 章/one/CountDownLatchOne.java
public class CountDownLatchOne {

    static final CountDownLatch startSignal = new CountDownLatch(1);

    public static void main(String[] args) throws Exception {
        new Thread(new Worker(),"A").start();
        System.out.println("doSomethingElse");
        Thread.sleep(2000);
        startSignal.countDown();
    }
```

```java
static class Worker implements Runnable {

    public void run() {
        try {
            startSignal.await();
            //使当前执行线程阻塞等待,直到计数器为 0 或者此线程中断
            System.out.println(Thread.currentThread().getName());
        } catch (InterruptedException ex) {
            ex.printStackTrace();
        }
    }

}
```

执行结果如下:

```
doSomethingElse
A
```

2. await(long timeout,TimeUnit unit)

使当前执行线程阻塞等待,直到计数器为 0、此线程中断或者超过最大等待时间,如果超过最大等待时间,则返回值为 false,否则返回值为 true。接收 long 入参,作为最大等待时间,接收 TimeUnit 入参,作为时间单位,代码如下:

```java
//第 9 章/one/CountDownLatchOne.java
public class CountDownLatchOne {

    static final CountDownLatch startSignal = new CountDownLatch(1);

    public static void main(String[] args) throws Exception {
        new Thread(new Worker(),"A").start();
        System.out.println("doSomethingElse");
        Thread.sleep(2000);
        startSignal.countDown();
    }

    static class Worker implements Runnable {

        public void run() {
            try {
                System.out.println(startSignal.await(1, TimeUnit.SECONDS));
                System.out.println(Thread.currentThread().getName());
            } catch (Exception ex) {
                ex.printStackTrace();
            }
        }
    }
}
```

执行结果如下：

```
doSomethingElse
false
A
```

3. countDown()

递减计数器，如果计数器达到0，则释放所有等待的线程。

4. getCount()

返回当前计数器，代码如下：

```java
//第9章/one/CountDownLatchTest.java
public class CountDownLatchTest {

    static final CountDownLatch startSignal = new CountDownLatch(1);
    static final CountDownLatch doneSignal = new CountDownLatch(10);

    public static void main(String[] args) throws Exception {
        for (int i = 0; i < 10; ++i) {
            new Thread(new Worker()).start();
        }
        System.out.println("doSomethingElse");
        Thread.sleep(2000);
        startSignal.countDown();
        doneSignal.await();
        System.out.println("main:" + doneSignal.getCount());
    }

    static class Worker implements Runnable {

        public void run() {
            try {
                startSignal.await();
                System.out.println(Thread.currentThread().getName());
                doWork();
            } catch (InterruptedException ex) {
                ex.printStackTrace();
            }
        }
        //类锁
        synchronized static void doWork() {
            System.out.println(doneSignal.getCount());
            doneSignal.countDown();
        }
    }
}
```

执行结果如下：

```
doSomethingElse
Thread - 2
Thread - 1
Thread - 0
Thread - 3
Thread - 4
Thread - 7
10
Thread - 8
Thread - 9
Thread - 6
Thread - 5
9
8
7
6
5
4
3
2
1
main:0
```

9.2 CyclicBarrier

一种同步辅助工具,它接受一个初始计数器,并允许多个执行线程等待,直到此计数器为 0 时唤醒所有等待的线程。此类称为循环屏障,因为它可以重复使用。

23min

9.2.1 构造器

CyclicBarrier 构造器见表 9-2。

表 9-2 CyclicBarrier 构造器

构 造 器	说 明
CyclicBarrier(int parties)	构造新的对象,指定初始计数器
CyclicBarrier(int parties,Runnable barrierAction)	构造新的对象,指定初始计数器,指定屏障完成时的回调实现类

9.2.2 常用方法

1. await()

使当前执行线程阻塞等待,直到计数器为 0、此线程中断或者任务被破坏。返回当前执行线程到达的屏障索引,代码如下:

```java
//第9章/two/CyclicBarrierTest.java
public class CyclicBarrierTest {

    private static final CyclicBarrier CYCLIC_BARRIER =
    new CyclicBarrier(3, new Runnable() {
        @Override
        public void run() {
            System.out.println("成团");
        }
    });
    public static void main(String[] args) throws Exception {
        RunnableImpl runnable = new RunnableImpl();
        new Thread(runnable,"A").start();
        new Thread(runnable,"B").start();
        Thread thread = new Thread(runnable, "C");
        thread.start();
    }

    static class RunnableImpl implements Runnable{

        @Override
        public void run() {
            try {
                CYCLIC_BARRIER.await();
                System.out.println(Thread.currentThread().getName());
            } catch (Exception e) {
                e.printStackTrace();
            }
        }
    }
}
```

执行结果如下：

```
成团
C
B
A
```

注意：此对象的 await() 方法会递减计数器。

2. await(long timeout，TimeUnit unit)

使当前执行线程阻塞等待，直到计数器为 0、此线程中断、任务被破坏或者超过最大等待时间，返回当前执行线程到达的屏障索引。接收 long 入参，作为最大等待时间，接收 TimeUnit 入参，作为时间单位，代码如下：

```java
//第9章/two/CyclicBarrierTest.java
public class CyclicBarrierTest {

    private static final CyclicBarrier CYCLIC_BARRIER =
    new CyclicBarrier(3, new Runnable() {
        @Override
        public void run() {
            System.out.println("成团");
        }
    });

    public static void main(String[] args) throws Exception {
        RunnableImpl runnable = new RunnableImpl();
        new Thread(runnable,"A").start();
        new Thread(runnable,"B").start();
        Thread.sleep(2000);
        Thread thread = new Thread(runnable, "C");
        thread.start();
    }

    static class RunnableImpl implements Runnable{

        @Override
        public void run() {
            try {
                if(Thread.currentThread().getName().equals("B")){
                    CYCLIC_BARRIER.await(1, TimeUnit.SECONDS);
                }else{
                    CYCLIC_BARRIER.await();
                }
                System.out.println(Thread.currentThread().getName());
            } catch (Exception e) {
                e.printStackTrace();
            }
        }
    }
}
```

执行结果如下：

```
java.util.concurrent.TimeoutException
    at
java.base/java.util.concurrent.CyclicBarrier.dowait(CyclicBarrier.java:259)
    at
java.base/java.util.concurrent.CyclicBarrier.await(CyclicBarrier.java:437)
    at
cn.kungreat.book.nine.two.CyclicBarrierTest$RunnableImpl.run(CyclicBarrierTest.java:30)
    at
java.base/java.lang.Thread.run(Thread.java:833)
```

```
java.util.concurrent.BrokenBarrierException
    at 
java.base/java.util.concurrent.CyclicBarrier.dowait(CyclicBarrier.java:252)
    at 
java.base/java.util.concurrent.CyclicBarrier.await(CyclicBarrier.java:364)
    at 
cn.kungreat.book.nine.two.CyclicBarrierTest $ RunnableImpl.run(CyclicBarrierTest.java:32)
    at 
java.base/java.lang.Thread.run(Thread.java:833)

java.util.concurrent.BrokenBarrierException
    at 
java.base/java.util.concurrent.CyclicBarrier.dowait(CyclicBarrier.java:210)
    at 
java.base/java.util.concurrent.CyclicBarrier.await(CyclicBarrier.java:364)
    at 
cn.kungreat.book.nine.two.CyclicBarrierTest $ RunnableImpl.run(CyclicBarrierTest.java:32)
    at 
java.base/java.lang.Thread.run(Thread.java:833)
```

注意：超时会影响本次屏障内的所有线程，一次屏障要么全部成功，要么全部失败。

3. getNumberWaiting()

返回当前在屏障处阻塞等待的线程数量。

```
public int getParties() {
    return parties;
}
```

图 9-1　getParties()方法源代码

4. getParties()

返回触发此屏障，完成所需要的参与方数量。构造器时传入的值，如图 9-1 所示。

5. isBroken()

返回此屏障是否处于损坏状态。

6. reset()

将屏障置为损坏状态，然后将屏障重置为初始状态。

9.3　Semaphore

此类被称为信号量或者许可证，此类提供了限制多线程并发数量的功能。

9.3.1　构造器

Semaphore 构造器见表 9-3。

表 9-3　Semaphore 构造器

构　造　器	说　　明
Semaphore(int permits)	构造新的对象，指定许可证数量
Semaphore(int permits, boolean fair)	构造新的对象，指定许可证数量，指定是否使用公平模式

9.3.2 常用方法

1. acquire()

阻塞等待获得许可证,直到获得可用许可证或者线程中断,代码如下:

```java
//第9章/three/SemaphoreTest.java
public class SemaphoreTest {

    private static final Semaphore SEMAPHORE = new Semaphore(2);

    public static void main(String[] args) throws Exception {
        RunnableImpl runnable = new RunnableImpl();
        new Thread(runnable,"A").start();
        new Thread(runnable,"B").start();
        new Thread(runnable, "C").start();
    }

    static class RunnableImpl implements Runnable{

        @Override
        public void run() {

                try {
                    SEMAPHORE.acquire();
                    System.out.println(Thread.currentThread().getName());
                    Thread.sleep(500);
                } catch (Exception e) {
                    e.printStackTrace();
                }

        }
    }
}
```

执行结果如下:

```
A
B
```

注意:A、B、C 这 3 个执行线程中会有一个得不到许可证,C 线程的概率最大。得不到许可证的执行线程将会一直阻塞等待。

2. acquire(int permits)

阻塞等待获得指定数量的许可证,直到获得指定数量的许可证或者线程中断。接收 int 入参,作为指定数量的许可证。

3. acquireUninterruptibly()

阻塞等待获得许可证,直到获得可用许可证。

4. acquireUninterruptibly(int permits)

阻塞等待获得指定数量的许可证,直到获得指定数量的许可证。接收 int 入参,作为指定数量的许可证。

5. availablePermits()

返回当前可用的许可证数量,代码如下:

```java
//第 9 章/three/SemaphoreTest.java
public class SemaphoreTest {

    private static final Semaphore SEMAPHORE = new Semaphore(2);

    public static void main(String[] args) throws Exception {
        System.out.println(SEMAPHORE.availablePermits());
    }

}
```

执行结果如下:

```
2
```

6. drainPermits()

回收当前可用许可证,并返回回收的数量,代码如下:

```java
//第 9 章/three/SemaphoreTest.java
public class SemaphoreTest {

    private static final Semaphore SEMAPHORE = new Semaphore(2);

    public static void main(String[] args) throws Exception {
        System.out.println(SEMAPHORE.drainPermits());
        System.out.println(SEMAPHORE.availablePermits());
    }

}
```

执行结果如下:

```
2
0
```

7. getQueuedThreads()

返回当前正在等待获得许可证的执行线程集合。

8. getQueueLength()

返回当前正在等待获得许可证的执行线程数量。

9. hasQueuedThreads()

返回当前是否有执行线程正在等待获得许可证。

10. isFair()

返回当前是否使用的是公平模式。

11. release()

归还许可证，代码如下：

```java
//第9章/three/SemaphoreTest.java
public class SemaphoreTest {

    private static final Semaphore SEMAPHORE = new Semaphore(2);

    public static void main(String[] args) throws Exception {
        RunnableImpl runnable = new RunnableImpl();
        new Thread(runnable,"A").start();
        new Thread(runnable,"B").start();
        new Thread(runnable, "C").start();
    }

    static class RunnableImpl implements Runnable{

        @Override
        public void run() {
                try {
                    SEMAPHORE.acquire();
                    System.out.println(Thread.currentThread().getName());
                    Thread.sleep(500);
                } catch (Exception e) {
                    e.printStackTrace();
                }finally {
                    SEMAPHORE.release(); //归还许可证
                }

        }
    }
}
```

执行结果如下：

```
B
A
C
```

注意：归还许可证会使当前许可证数量增加，在没有获得许可证的情况下就归还许可证，会导致许可证数量超过构造对象时传入的值。

12. release(int permits)

归还指定数量的许可证。接收 int 入参，作为指定数量的许可证。

13. tryAcquire()

尝试获得许可证并立即返回，如果成功，则返回值为 true，如果失败，则返回值为 false，代码如下：

```java
//第9章/three/SemaphoreTest.java
public class SemaphoreTest {

    private static final Semaphore SEMAPHORE = new Semaphore(2);

    public static void main(String[] args) throws Exception {
        RunnableImpl runnable = new RunnableImpl();
        new Thread(runnable,"A").start();
        Thread.sleep(100);
        new Thread(runnable,"B").start();
        new Thread(runnable, "C").start();
        new Thread(runnable,"D").start();
    }

    static class RunnableImpl implements Runnable{

        @Override
        public void run() {
            if(SEMAPHORE.tryAcquire()){
                try {
                    System.out.println(Thread.currentThread().getName());
                    Thread.sleep(500);
                } catch (Exception e) {
                    e.printStackTrace();
                }finally {
                    SEMAPHORE.release();
                }
            }else{
                System.out.println("no Semaphore");
            }
        }
    }
}
```

执行结果如下：

```
A
B
no Semaphore
no Semaphore
```

14. tryAcquire(int permits)

尝试获得指定数量的许可证并立即返回，如果成功，则返回值为 true，如果失败，则返回值为 false。接收 int 入参，作为指定数量的许可证。

15. tryAcquire(long timeout,TimeUnit unit)

在最大等待时间内尝试获得许可证，如果成功，则返回值为 true，如果失败，则返回值为 false。接收 long 入参，作为最大等待时间，接收 TimeUnit 入参，作为时间单位，代码如下：

```java
//第9章/three/SemaphoreTest.java
public class SemaphoreTest {

    private static final Semaphore SEMAPHORE = new Semaphore(2);

    public static void main(String[] args) throws Exception {
        RunnableImpl runnable = new RunnableImpl();
        new Thread(runnable,"A").start();
        new Thread(runnable,"B").start();
        new Thread(runnable, "C").start();
        new Thread(runnable,"D").start();
    }

    static class RunnableImpl implements Runnable{
        @Override
        public void run() {
            try {
                if(SEMAPHORE.tryAcquire(1, TimeUnit.SECONDS)){
                    try {
                        System.out.println(Thread.currentThread().getName());
                        Thread.sleep(1500);
                    } catch (Exception e) {
                        e.printStackTrace();
                    }finally {
                        SEMAPHORE.release();
                    }
                }else{
                    System.out.println("no Semaphore");
                }
            } catch (InterruptedException e) {
                e.printStackTrace();
            }
        }
    }
}
```

执行结果如下：

```
B
A
no Semaphore
no Semaphore
```

9.4 Phaser

37min

可重用的同步屏障,在功能上类似于CyclicBarrier、CountDownLatch但支持更灵活的使用方式。

9.4.1 构造器

25min

Phaser构造器见表9-4。

表9-4 Phaser构造器

构造器	说明
Phaser()	构造新的对象,默认为无参构造器
Phaser(int parties)	构造新的对象,指定初始计数器
Phaser(Phaser parent)	构造新的对象,指定上级对象
Phaser(Phaser parent, int parties)	构造新的对象,指定上级对象,指定初始计数器

9.4.2 常用方法

18min

1. arrive()

到达这个分阶器并使当前计数器递减,无须阻塞等待其他执行线程到达。返回已经完成的分阶器次数,如果终止,则为负值,代码如下:

```java
//第9章/four/PhaserTest.java
public class PhaserTest {

    private static final Phaser PHASER = new Phaser(2);

    public static void main(String[] args) {
        System.out.println(PHASER.getRegisteredParties());
                                            //当前总计数器数量
        System.out.println(PHASER.getArrivedParties());
                                            //当前已到达计数器数量
        System.out.println(PHASER.getUnarrivedParties());
                                            //当前未到达计数器数量
        System.out.println(PHASER.arrive());
        System.out.println(PHASER.getRegisteredParties());
                                            //当前总计数器数量
        System.out.println(PHASER.getArrivedParties());
                                            //当前已到达计数器数量
        System.out.println(PHASER.getUnarrivedParties());
                                            //当前未到达计数器数量
    }
}
```

执行结果如下:

```
2
0
2
0
2
1
1
```

2. arriveAndAwaitAdvance()

到达这个分阶器并使当前计数器递减,阻塞等待其他执行线程到达。返回已经完成的分阶器次数,如果终止,则为负值,代码如下:

```java
public class PhaserTest {

    private static final Phaser PHASER = new Phaser(2);

    public static void main(String[] args) {
        System.out.println(PHASER.getRegisteredParties());
                                                    //当前总计数器数量
        System.out.println(PHASER.getArrivedParties());
                                                    //当前已到达计数器数量
        System.out.println(PHASER.getUnarrivedParties());
                                                    //当前未到达计数器数量
        System.out.println(PHASER.arriveAndAwaitAdvance());
        System.out.println(PHASER.getRegisteredParties());
                                                    //当前总计数器数量
        System.out.println(PHASER.getArrivedParties());
                                                    //当前已到达计数器数量
        System.out.println(PHASER.getUnarrivedParties());
                                                    //当前未到达计数器数量
    }
}
```

执行结果如下:

```
2
0
2
```

注意:主执行线程调用 arriveAndAwaitAdvance() 方法后阻塞等待其他执行线程到达。

修改 PhaserTest 类,代码如下:

```java
public class PhaserTest {

    private static final Phaser PHASER = new Phaser(2);

    public static void main(String[] args) {
```

```java
            System.out.println(PHASER.getRegisteredParties());
                                                    //当前总计数器数量
            System.out.println(PHASER.getArrivedParties());
                                                    //当前已到达计数器数量
            System.out.println(PHASER.getUnarrivedParties());
                                                    //当前未到达计数器数量
            new Thread(new Runnable() {
                @Override
                public void run() {
                    System.out.println("arriveAndAwaitAdvance:"
                                    + PHASER.arriveAndAwaitAdvance());
                }
            }).start();
            System.out.println("arriveAndAwaitAdvance:"
                                    + PHASER.arriveAndAwaitAdvance());
            System.out.println(PHASER.getRegisteredParties());
                                                    //当前总计数器数量
            System.out.println(PHASER.getArrivedParties());
                                                    //当前已到达计数器数量
            System.out.println(PHASER.getUnarrivedParties());
                                                    //当前未到达计数器数量
        }
    }
```

执行结果如下:

```
2
0
2
arriveAndAwaitAdvance:1
arriveAndAwaitAdvance:1
2
0
2
```

3. arriveAndDeregister()

到达这个分阶器并使当前计数器取消注册一个,无须阻塞等待其他执行线程到达。返回已经完成的分阶器次数,如果终止,则为负值,代码如下:

```java
//第9章/four/PhaserTest.java
public class PhaserTest {

    private static final Phaser PHASER = new Phaser(2);

    public static void main(String[] args) {
        System.out.println(PHASER.getRegisteredParties());
                                                    //当前总计数器数量
        System.out.println(PHASER.getArrivedParties());
                                                    //当前已到达计数器数量
```

```
            System.out.println(PHASER.getUnarrivedParties());
                                            //当前未到达计数器数量
            System.out.println("arriveAndDeregister:"
                            + PHASER.arriveAndDeregister());
            System.out.println("arriveAndAwaitAdvance:"
                            + PHASER.arriveAndAwaitAdvance());
            System.out.println(PHASER.getRegisteredParties());
                                            //当前总计数器数量
            System.out.println(PHASER.getArrivedParties());
                                            //当前已到达计数器数量
            System.out.println(PHASER.getUnarrivedParties());
                                            //当前未到达计数器数量
    }
}
```

执行结果如下：

```
2
0
2
arriveAndDeregister:0
arriveAndAwaitAdvance:1
1
0
1
```

注意：此分阶器的当前总计数器数量的最后输出为1。

4. bulkRegister(int parties)

将给定数量的计数器添加到此分阶器中，返回已经完成的分阶器次数。如果终止，则为负值，在这种情况下注册无效。接收 int 入参，作为指定数量的计数器，代码如下：

```java
//第 9 章/four/PhaserTest.java
public class PhaserTest {

    private static final Phaser PHASER = new Phaser(2);

    public static void main(String[] args) {
        System.out.println(PHASER.getRegisteredParties());
                                            //当前总计数器数量
        System.out.println("bulkRegister:" + PHASER.bulkRegister(3));
        System.out.println(PHASER.getRegisteredParties());
                                            //当前总计数器数量
    }
}
```

执行结果如下：

```
2
bulkRegister:0
5
```

5. forceTermination()

强制此相位器进入终止状态，代码如下：

```java
//第 9 章/four/PhaserTest.java
public class PhaserTest {

    private static final Phaser PHASER = new Phaser(5);

    public static void main(String[] args) throws Exception {
        for (int i = 0; i < 4; i++) {
            new Thread(new Runnable() {
                @Override
                public void run() {
                    PHASER.arriveAndAwaitAdvance();
                    if(PHASER.isTerminated()){
                        System.out.println("此分阶器进入终止状态");
                    }
                }
            }).start();
        }
        Thread.sleep(1000);
        PHASER.forceTermination(); //强制此相位器进入终止状态
    }
}
```

执行结果如下：

```
此分阶器进入终止状态
此分阶器进入终止状态
此分阶器进入终止状态
此分阶器进入终止状态
```

6. getArrivedParties()

返回当前已到达计数器数量。

7. getPhase()

返回已经完成的分阶器次数，代码如下：

```java
//第 9 章/four/PhaserTest.java
public class PhaserTest {

    private static final Phaser PHASER = new Phaser(5);

    public static void main(String[] args) throws Exception {
```

```
            for (int i = 0; i < 15; i++) {
                new Thread(new Runnable() {
                    @Override
                    public void run() {
                        PHASER.arriveAndAwaitAdvance();
                    }
                }).start();
            }
            Thread.sleep(2000);
            System.out.println(PHASER.getPhase());
    }
}
```

执行结果如下：

```
3
```

8. getRegisteredParties()

返回当前总计数器数量。

9. getRoot()

返回此分阶器的上级，如果它没有上级，则返回此分阶器。

10. getUnarrivedParties()

返回当前未到达计数器数量。

11. isTerminated()

返回此分阶器是否进入终止状态。

12. register()

将计数器递增，返回已经完成的分阶器次数。如果终止，则为负值，在这种情况下注册无效，代码如下：

```java
//第9章/four/PhaserTest.java
public class PhaserTest {

    private static final Phaser PHASER = new Phaser(2);

    public static void main(String[] args) throws Exception {
        System.out.println(PHASER.getRegisteredParties());
        System.out.println(PHASER.register());
        System.out.println(PHASER.getRegisteredParties());
    }
}
```

执行结果如下：

```
2
0
3
```

13. onAdvance(int phase, int registeredParties)

可重写的回调方法，用于在即将发生的分阶器完成事件时执行回调，并控制是否终止此分阶器，代码如下：

```java
//第9章/four/PhaserTest.java
public class PhaserTest {

    private static final Phaser PHASER = new MyPhaser(5);

    public static void main(String[] args) throws Exception {
        for (int i = 0; i < 5; i++) {
            new Thread(new Runnable() {
                @Override
                public void run() {
                    PHASER.arriveAndAwaitAdvance();
                }
            }).start();
        }
        Thread.sleep(2000);
        for (int i = 0; i < 5; i++) {
            new Thread(new Runnable() {
                @Override
                public void run() {
                    PHASER.arriveAndAwaitAdvance();
                }
            }).start();
        }
        Thread.sleep(2000);
        for (int i = 0; i < 5; i++) {
            new Thread(new Runnable() {
                @Override
                public void run() {
                    PHASER.arriveAndAwaitAdvance();
                }
            }).start();
        }
        Thread.sleep(2000);
        System.out.println("isTerminated:1 - " + PHASER.isTerminated());
        for (int i = 0; i < 5; i++) {
            new Thread(new Runnable() {
                @Override
                public void run() {
                    PHASER.arriveAndAwaitAdvance();
                }
            }).start();
        }
        Thread.sleep(2000);
        System.out.println("isTerminated:2 - " + PHASER.isTerminated());
    }
```

```java
static final class MyPhaser extends Phaser{

    public MyPhaser(int parties) {
        super(parties);
    }

    @Override
    protected boolean onAdvance(int phase, int registeredParties) {
        System.out.println("registeredParties:" + registeredParties);
        switch (phase) {
            case 0 -> System.out.println("开始");
            case 1 -> System.out.println("1 阶段");
            case 2 -> System.out.println("2 阶段");
            case 3 -> {
                System.out.println("终止阶段");
                return true;
            }
        }
        return false;
    }
}
```

执行结果如下：

```
registeredParties:5
开始
registeredParties:5
1 阶段
registeredParties:5
2 阶段
isTerminated:1 - false
registeredParties:5
终止阶段
isTerminated:2 - true
```

14. awaitAdvance(int phase)

等待此分阶器完成指定的次数，如果当前完成的次数不等于给定的值或者此分阶器终止，则立即返回。返回下一个完成的次数。接收 int 入参，作为指定的次数，代码如下：

```java
//第 9 章/four/PhaserTestTwo.java
public class PhaserTestTwo {

    private static final Phaser PHASER = new Phaser(5);

    public static void main(String[] args) throws Exception {

        new Thread(new Runnable() {
            @Override
```

```java
            public void run() {
                System.out.println("awaitAdvance:" + PHASER.awaitAdvance(0));
            }
        }).start();

        for (int i = 0; i < 5; i++) {
            new Thread(new Runnable() {
                @Override
                public void run() {
                    try {
                        Thread.sleep(1000);
                        PHASER.arriveAndAwaitAdvance();
                    } catch (InterruptedException e) {
                        e.printStackTrace();
                    }
                }
            }).start();
        }
    }
}
```

执行结果如下：

```
awaitAdvance:1
```

小结

官方提供的同步器使用起来还是比较方便的，如果可以，则读者应自己去实现一些同步器工具类，合理地运用以前所学的知识点就已经足够了。

习题

1. 选择题

（1）CountDownLatch 的（　　）方法可以使当前执行线程阻塞等待。（多选）

　　A. await()　　　　　　　　　　　　　　B. countDown()

　　C. await(long timeout, TimeUnit unit)　　D. getCount()

（2）CyclicBarrier 的（　　）方法有重置效果。（单选）

　　A. getParties()　　B. isBroken()　　C. reset()　　D. await()

2. 填空题

（1）根据业务要求补全代码，主执行线程阻塞等待其他执行线程完成任务，代码如下：

```java
//第 9 章/answer/OneAnswer.java
public class OneAnswer {
```

```java
    static final CountDownLatch COUNT = new CountDownLatch(_____);

    public static void main(String[] args) throws Exception {
        RunnableImpl runnable = new RunnableImpl();
        new Thread(runnable,"A").start();
        new Thread(runnable,"B").start();
        _____;
        System.out.println("main - end");
    }

    static final class RunnableImpl implements Runnable{

        @Override
        public void run() {
            _____;
        }
    }
}
```

（2）根据业务要求补全代码，5 个执行线程阻塞等待，直到主执行线程睡眠 2s 后再唤醒 5 个阻塞等待的执行线程，代码如下：

```java
//第 9 章/answer/TwoAnswer.java
public class TwoAnswer {
    static final CountDownLatch COUNT = new CountDownLatch(_____);

    public static void main(String[] args) throws Exception {
        RunnableImpl runnable = new RunnableImpl();
        for (int i = 0; i < 5; i++) {
            new Thread(runnable).start();
        }
        Thread.sleep(2000);
        _____;
        System.out.println("main - end");
    }

    static final class RunnableImpl implements Runnable{

        @Override
        public void run() {
            try {
                _____;
                System.out.println(Thread.currentThread().getName());
            } catch (InterruptedException e) {
                e.printStackTrace();
            }
        }
    }
}
```

（3）根据业务要求补全代码，将多线程并发数量控制为10，代码如下：

```java
//第 9 章/answer/ThreeAnswer.java
public class ThreeAnswer {
    static final Semaphore SEMAPHORE = new Semaphore(_____);

    public static void main(String[] args) throws Exception {
        RunnableImpl runnable = new RunnableImpl();
        for (int i = 0; i < 50; i++) {
            new Thread(runnable).start();
        }
    }

    static final class RunnableImpl implements Runnable {

        @Override
        public void run() {
            try {
                SEMAPHORE._____;
                Thread.sleep(2000);
                System.out.println(Thread.currentThread().getName());
            } catch (Exception e) {
                e.printStackTrace();
            } finally{
                SEMAPHORE._____;
            }
        }
    }
}
```

第 10 章 AQS 源码分析

提供了一个框架,用于实现先进先出(FIFO)阻塞等待链表和相关的锁功能。此类被设计为大多数锁、同步器的基础实现类。例如 Lock、CountDownLatch、Semaphore 等实现类的功能设计都依赖了 AbstractQueuedSynchronizer。

10.1 构造器

AbstractQueuedSynchronizer 构造器见表 10-1。

表 10-1 AbstractQueuedSynchronizer 构造器

构造器	说明
AbstractQueuedSynchronizer()	构造新的对象,默认为无参构造器

官方文档示例非重入互斥锁,代码如下:

```java
//第 10 章/one/MyLock.java
public class MyLock {
    private final Sync sync = new Sync();

    private static class Sync extends AbstractQueuedSynchronizer {
        //加锁时回调此方法
        public boolean tryAcquire(int acquires) {
            if (compareAndSetState(0, 1)) {
                setExclusiveOwnerThread(Thread.currentThread());
                return true;
            }
            return false;
        }
        //释放锁时回调此方法
        protected boolean tryRelease(int releases) {
            if (!isHeldExclusively())
                throw new IllegalMonitorStateException();
            setExclusiveOwnerThread(null);
            setState(0);
            return true;
```

```
        }
        //是否是当前执行线程对象拥有此锁
        public boolean isHeldExclusively() {
            return getExclusiveOwnerThread() == Thread.currentThread();
        }

        //是否有执行线程对象拥有此锁
        public boolean isLocked() {
            return getState() != 0;
        }

    }

    public void lock()            { sync.acquire(1); }
    public boolean tryLock()      { return sync.tryAcquire(1); }
    public void unlock()          { sync.release(1); }
    public boolean isLocked()     { return sync.isLocked(); }
    public boolean isHeldByCurrentThread() {
        return sync.isHeldExclusively();
    }
    public void lockInterruptibly() throws InterruptedException {
        sync.acquireInterruptibly(1);
    }
    public boolean tryLock(long timeout, TimeUnit unit)
            throws InterruptedException {
        return sync.tryAcquireNanos(1, unit.toNanos(timeout));
    }
}
```

10.2 常用方法

134min

1. acquire(int arg)

以独占模式获得锁忽略线程中断,会回调 tryAcquire(int arg)方法,此方法的返回值会影响最终的执行结果,如果拿锁失败,则会造成当前执行线程阻塞等待。接收 int 入参,作为同步状态数量,代码如下:

```
//第 10 章/one/AQSTest.java
public class AQSTest {
    private static final Sync SYNC = new Sync();

    public static void main(String[] args) throws Exception {
        Thread thread = new Thread(new Runnable() {
            @Override
            public void run() {
                SYNC.acquire(1);
                try {
```

```java
                    Thread.sleep(2000);
                } catch (Exception e) {
                    e.printStackTrace();
                } finally {
                    SYNC.release(1);
                }
            }
        }, "A");
        thread.start();

        SYNC.acquire(1);
        try {
            Thread.sleep(3000);
        }finally {
            SYNC.release(1);
        }
    }

    private static class Sync extends AbstractQueuedSynchronizer {

        //加锁时回调此方法
        public boolean tryAcquire(int acquires) {
            if (compareAndSetState(0, 1)) {
                setExclusiveOwnerThread(Thread.currentThread());
                System.out.println(Thread.currentThread().getName()
                                                      + "拿 Lock");

                return true;
            }
            return false;
        }
        //释放锁时回调此方法
        protected boolean tryRelease(int releases) {
            if (!isHeldExclusively())
                throw new IllegalMonitorStateException();
            setExclusiveOwnerThread(null);
            setState(0);
            System.out.println(Thread.currentThread().getName()
                                                + "释放 Lock");
            return true;
        }
        //是否是当前执行线程对象拥有此锁
        public boolean isHeldExclusively() {
            return getExclusiveOwnerThread() == Thread.currentThread();
        }

        //是否有执行线程对象拥有此锁
        public boolean isLocked() {
            return getState() != 0;
        }
    }
}
```

执行结果如下:

```
main 拿 Lock
main 释放 Lock
A 拿 Lock
A 释放 Lock
```

2. acquireInterruptibly(int arg)

以独占模式获得锁响应线程中断,会回调 tryAcquire(int arg)方法,此方法的返回值会影响最终的执行结果,拿锁失败会造成当前执行线程阻塞等待。接收 int 入参,作为同步状态数量,代码如下:

```java
//第 10 章/one/AQSTest.java
public class AQSTest {
    private static final Sync SYNC = new Sync();

    public static void main(String[] args) throws InterruptedException {
        Thread thread = new Thread(new Runnable() {
            @Override
            public void run() {
                try {
                    Thread.sleep(200);
                    SYNC.acquireInterruptibly(1);
                } catch (InterruptedException e) {
                    e.printStackTrace();
                } finally {
                    if(SYNC.isHeldExclusively()){
                        SYNC.release(1);
                    }
                }
            }
        }, "A");
        thread.start();

        SYNC.acquire(1);
        try {
            Thread.sleep(1000);
            thread.interrupt();
        }finally {
            SYNC.release(1);
        }
    }

    private static class Sync extends AbstractQueuedSynchronizer {

        //加锁时回调此方法
        public boolean tryAcquire(int acquires) {
            if (compareAndSetState(0, 1)) {
                setExclusiveOwnerThread(Thread.currentThread());
```

```
                System.out.println(Thread.currentThread().getName()
                                        + "拿 Lock");
                return true;
            }
            return false;
        }
        //释放锁时回调此方法
        protected boolean tryRelease(int releases) {
            if (!isHeldExclusively())
                throw new IllegalMonitorStateException();
            setExclusiveOwnerThread(null);
            setState(0);
            System.out.println(Thread.currentThread().getName()
                                        + "释放 Lock");
            return true;
        }
        //是否是当前执行线程对象拥有此锁
        public boolean isHeldExclusively() {
            return getExclusiveOwnerThread() == Thread.currentThread();
        }

        //是否有执行线程对象拥有此锁
        public boolean isLocked() {
            return getState() != 0;
        }
    }
}
```

执行结果如下：

```
main 拿 Lock
main 释放 Lock
java.lang.InterruptedException
    at java.base/java.util.concurrent.locks.AbstractQueuedSynchronizer.
acquireInterruptibly(AbstractQueuedSynchronizer.java:959)
    at cn.kungreat.book.ten.one.AQSTest$1.run(AQSTest.java:14)
    at java.base/java.lang.Thread.run(Thread.java:833)
```

3. release(int arg)

以独占模式释放锁。会回调 tryRelease(int releases)方法，此方法的返回值会影响最终的执行结果，如果成功，则会唤醒链表上指定节点的线程并返回 true，如果失败，则返回 false。接收 int 入参，作为同步状态数量。

4. acquireShared(int arg)

以共享模式获得锁忽略线程中断，会回调 tryAcquireShared(int acquires)方法，此方法的返回值会影响最终的执行结果，如果拿锁失败，则会造成当前执行线程阻塞等待。接收 int 入参，作为同步状态数量，代码如下：

```java
//第10章/one/AQSShareTest.java
public class AQSShareTest {

    private static final Sync sync = new Sync(5);

    public static void main(String[] args) {
        for (int i = 0; i < 5; i++) {
            new Thread(new Runnable() {
                @Override
                public void run() {
                    sync.releaseShared(1);
                    System.out.println(Thread.currentThread().getName()
                            + ":end");
                }
            }).start();
        }
        sync.acquireShared(1);
        System.out.println("main-end");
    }

    private static final class Sync extends AbstractQueuedSynchronizer {

        Sync(int count) {
            setState(count);
        }

        int getCount() {
            return getState();
        }

        protected int tryAcquireShared(int acquires) {
            return (getState() == 0) ? 1 : -1;
        }

        protected boolean tryReleaseShared(int releases) {
            for (;;) {
                int c = getState();
                if (c == 0)
                    return false;
                int nextc = c - 1;
                if (compareAndSetState(c, nextc))
                    return nextc == 0;
            }
        }
    }
}
```

执行结果如下：

```
main-end
Thread-4:end
```

```
Thread-1:end
Thread-3:end
Thread-2:end
Thread-0:end
```

5. acquireSharedInterruptibly(int arg)

以共享模式获得锁响应线程中断，会回调 tryAcquireShared(int acquires)方法，此方法的返回值会影响最终的执行结果，如果拿锁失败，则会造成当前执行线程阻塞等待。接收 int 入参，作为同步状态数量。

6. releaseShared(int arg)

以共享模式释放锁。会回调 tryReleaseShared(int releases)方法，此方法的返回值会影响最终的执行结果，如果成功，则会唤醒链表上指定节点的线程并返回 true，如果失败，则返回 false。接收 int 入参，作为同步状态数量。

注意：共享模式的唤醒具有迭代效果。

7. getExclusiveQueuedThreads()

返回一个集合，其中包含正在等待以独占模式获得锁的线程对象，如图 10-1 所示。

```java
public final Collection<Thread> getExclusiveQueuedThreads() {
    ArrayList<Thread> list = new ArrayList<>();
    for (Node p = tail; p != null; p = p.prev) {
        if (!(p instanceof SharedNode)) {  // 过滤掉共享节点
            Thread t = p.waiter;
            if (t != null)
                list.add(t);
        }
    }
    return list;
}
```

图 10-1　getExclusiveQueuedThreads()方法源代码

8. getFirstQueuedThread()

返回链表中的第 1 个（等待时间最长）线程对象，如果没有，则返回空，如图 10-2 所示。

9. getQueuedThreads()

返回一个集合，其中包含正在等待获得锁的线程对象，如图 10-3 所示。

10. getQueueLength()

返回正在等待获得锁的线程对象的总数量，如图 10-4 所示。

11. getSharedQueuedThreads()

返回一个集合，其中包含正在等待以共享模式获得锁的线程对象，如图 10-5 所示。

12. hasContended()

查询链表节点是否有初始化，返回 boolean 值，如图 10-6 所示。

```java
public final Thread getFirstQueuedThread() {
    Thread first = null, w; Node h, s;
    if ((h = head) != null && ((s = h.next) == null ||
                               (first = s.waiter) == null ||
                               s.prev == null)) {

        for (Node p = tail, q; p != null && (q = p.prev) != null; p = q)
            if ((w = p.waiter) != null)
                first = w;
    }
    return first;
}
```

图 10-2　getFirstQueuedThread()方法源代码

```java
public final Collection<Thread> getQueuedThreads() {
    ArrayList<Thread> list = new ArrayList<>();
    for (Node p = tail; p != null; p = p.prev) {
        Thread t = p.waiter;
        if (t != null)
            list.add(t);
    }
    return list;
}
```

图 10-3　getQueuedThreads()方法源代码

```java
public final int getQueueLength() {
    int n = 0;
    for (Node p = tail; p != null; p = p.prev) {
        if (p.waiter != null)
            ++n;
    }
    return n;
}
```

图 10-4　getQueueLength()方法源代码

```java
public final Collection<Thread> getSharedQueuedThreads() {
    ArrayList<Thread> list = new ArrayList<>();
    for (Node p = tail; p != null; p = p.prev) {
        if (p instanceof SharedNode) {    // ← 共享模式节点
            Thread t = p.waiter;
            if (t != null)
                list.add(t);
        }
    }
    return list;
}
```

图 10-5　getSharedQueuedThreads()方法源代码

```java
public final boolean hasContended() {
    return head != null;
}
```

图 10-6　hasContended()方法源代码

13. hasQueuedPredecessors()

查询链表节点中是否有其他线程正在等待获得此锁，返回 boolean 值，如图 10-7 所示。

```java
public final boolean hasQueuedPredecessors() {
    Thread first = null; Node h, s;
    if ((h = head) != null && ((s = h.next) == null ||
                               (first = s.waiter) == null ||
                               s.prev == null))
        first = getFirstQueuedThread();
    return first != null && first != Thread.currentThread();
}
```

图 10-7　hasQueuedPredecessors()方法源代码

14. hasQueuedThreads()

查询链表节点中是否有其他节点，根据状态判断，返回 boolean 值，如图 10-8 所示。

```java
public final boolean hasQueuedThreads() {
    for (Node p = tail, h = head; p != h && p != null; p = p.prev)
        if (p.status >= 0)   ← 根据状态判断
            return true;
    return false;
}
```

图 10-8　hasQueuedThreads()方法源代码

15. isQueued(Thread thread)

如果给定线程对象当前已加入链表队列，则返回值为 true，否则返回值为 false。接收 Thread 入参，作为给定线程对象，如图 10-9 所示。

```java
public final boolean isQueued(Thread thread) {
    if (thread == null)
        throw new NullPointerException();
    for (Node p = tail; p != null; p = p.prev)
        if (p.waiter == thread)
            return true;
    return false;
}
```

图 10-9　isQueued()方法源代码

16. tryAcquireNanos(int arg, long nanosTimeout)

最大等待时间内以独占模式获得锁响应线程中断，会回调 tryAcquire(int arg) 方法，此方法的返回值会影响最终的执行结果,返回 boolean 值。接收 int 入参,作为同步状态数量,接收 long 入参,作为最大等待时间纳秒数,如图 10-10 所示。

```java
public final boolean tryAcquireNanos(int arg, long nanosTimeout)
        throws InterruptedException {
    if (!Thread.interrupted()) {
        if (tryAcquire(arg))
            return true;
        if (nanosTimeout <= 0L)
            return false;
        int stat = acquire( node: null, arg, shared: false, interruptible: true, timed: true,
                            time: System.nanoTime() + nanosTimeout);
        if (stat > 0)
            return true;
        if (stat == 0)
            return false;
    }
    throw new InterruptedException();
}
```

图 10-10　tryAcquireNanos()方法源代码

17. tryAcquireSharedNanos(int arg,long nanosTimeout)

最大等待时间内以共享模式获得锁响应线程中断,会回调 tryAcquireShared(int acquires)方法,此方法的返回值会影响最终的执行结果,返回 boolean 值。接收 int 入参,作为同步状态数量,接收 long 入参,作为最大等待时间纳秒数,如图 10-11 所示。

```java
public final boolean tryAcquireSharedNanos(int arg, long nanosTimeout)
        throws InterruptedException {
    if (!Thread.interrupted()) {
        if (tryAcquireShared(arg) >= 0)
            return true;
        if (nanosTimeout <= 0L)
            return false;
        int stat = acquire( node: null, arg, shared: true, interruptible: true, timed: true,
                            time: System.nanoTime() + nanosTimeout);
        if (stat > 0)
            return true;
        if (stat == 0)
            return false;
    }
    throw new InterruptedException();
}
```

图 10-11　tryAcquireSharedNanos()方法源代码

10.3 ConditionObject

此类是由非 static 修饰的内部类,默认会存在外部类对象地址引用,可以直接调用外部类对象的方法。此类实现了 Condition 接口,并且此类内部也维护了一套链表节点流程,核心字段如图 10-12 所示。

```
First node of condition queue.
private transient ConditionNode firstWaiter;
Last node of condition queue.
private transient ConditionNode lastWaiter;
```

图 10-12　核心字段

await() 方法会创建 ConditionNode 节点对象,并把此节点对象连接到 ConditionObject 对象里的链表节点中。最终执行线程会进入阻塞等待中,被唤醒后将汇入 AQS 主流程中,如图 10-13 所示。

```java
public final void await() throws InterruptedException {
    if (Thread.interrupted())
        throw new InterruptedException();
    ConditionNode node = new ConditionNode();// 创建节点
    int savedState = enableWait(node);// 连接节点链路
    LockSupport.setCurrentBlocker(this);
    boolean interrupted = false, cancelled = false, rejected = false;
    while (!canReacquire(node)) {// 是否汇入AQS主流程
        if (interrupted |= Thread.interrupted()) {
            if (cancelled = (node.getAndUnsetStatus(COND) & COND) != 0)
                break;
        } else if ((node.status & COND) != 0) {
            try {
                if (rejected)
                    node.block();// 阻塞等待
                else
                    ForkJoinPool.managedBlock(node);// 阻塞等待
            } catch (RejectedExecutionException ex) {
                rejected = true;
            } catch (InterruptedException ie) {
                interrupted = true;
            }
        } else
            Thread.onSpinWait();
    }
    LockSupport.setCurrentBlocker(null);
    node.clearStatus();// 清理节点的状态
    acquire(node, savedState, false, false, false, 0L);// 汇入AQS主流程
    if (interrupted) {
        if (cancelled) {
            unlinkCancelledWaiters(node);
            throw new InterruptedException();
        }
        Thread.currentThread().interrupt();
    }
}
```

图 10-13　await() 方法源代码

enableWait(ConditionNode node)方法用于检查当前执行线程是否是当前独占锁的拥有者并且释放当前独占锁。成功把此节点对象连接到 ConditionObject 对象里的链表节点中，如果失败，则抛出 IllegalMonitorStateException 异常，如图 10-14 所示。

signal()方法用于检查当前执行线程是否是当前独占锁的拥有者，如果成功，则把 ConditionObject 对象链表中的指定节点迁移到 AQS 链表节点的尾部，如果失败，则抛出 IllegalMonitorStateException 异常，如图 10-15 所示。

```java
private int enableWait(ConditionNode node) {
    if (isHeldExclusively()) {
        node.waiter = Thread.currentThread();
        node.setStatusRelaxed(COND | WAITING);
        ConditionNode last = lastWaiter;
        if (last == null)
            firstWaiter = node;
        else
            last.nextWaiter = node;
        lastWaiter = node;
        int savedState = getState();
        if (release(savedState))
            return savedState;
    }
    node.status = CANCELLED;
    throw new IllegalMonitorStateException();
}
```

图 10-14　enableWait()方法源代码

```java
public final void signal() {
    ConditionNode first = firstWaiter;
    if (!isHeldExclusively())
        throw new IllegalMonitorStateException();
    if (first != null)
        doSignal(first, all: false);
}
```

图 10-15　signal()方法源代码

doSignal(ConditionNode first, boolean all)方法用于清理掉 ConditionObject 对象链表中的指定节点，并把此节点添加到 AQS 链表节点的尾部，如图 10-16 所示。

```java
private void doSignal(ConditionNode first, boolean all) {
    while (first != null) {
        ConditionNode next = first.nextWaiter;
        if ((firstWaiter = next) == null)
            lastWaiter = null;                      // 清理节点
        if ((first.getAndUnsetStatus(COND) & COND) != 0) {
            enqueue(first);                         // 把此节点添加到AQS链表节点的尾部
            if (!all)
                break;
        }
        first = next;
    }
}
```

图 10-16　doSignal()方法源代码

enqueue(Node node)方法是 AQS 对象的方法，用于把指定节点添加到 AQS 链表节点的尾部，如图 10-17 所示。

```
final void enqueue(Node node) {
    if (node != null) {
        for (;;) {
            Node t = tail;
            node.setPrevRelaxed(t);    //连接上一个节点
            if (t == null)              //初始化
                tryInitializeHead();
            else if (casTail(t, node)) {//cas操作
                t.next = node;
                if (t.status < 0)       //唤醒以清理节点链路
                    LockSupport.unpark(node.waiter);
                break;
            }
        }
    }
}
```

图 10-17　enqueue()方法源代码

> 注意：同一个 AQS 对象，可以创建多个 ConditionObject 内部类对象，Lock 锁的 newCondition()方法就是基于此特性实现的。

小结

AbstractQueuedSynchronizer 提供了一套可用于实现同步锁机制的框架，需要去重写几个主要的方法，实现自己的锁机制。JDK 8、JDK 11、JDK 17 源代码的实现略有差异。

图 书 推 荐

书　名	作　者
深度探索 Vue.js——原理剖析与实战应用	张云鹏
剑指大前端全栈工程师	贾志杰、史广、赵东彦
Flink 原理深入与编程实战——Scala＋Java(微课视频版)	辛立伟
Spark 原理深入与编程实战(微课视频版)	辛立伟、张帆、张会娟
HarmonyOS 应用开发实战(JavaScript 版)	徐礼文
HarmonyOS 原子化服务卡片原理与实战	李洋
鸿蒙操作系统开发入门经典	徐礼文
鸿蒙应用程序开发	董昱
鸿蒙操作系统应用开发实践	陈美汝、郑森文、武延军、吴敬征
HarmonyOS 移动应用开发	刘安战、余雨萍、李勇军 等
HarmonyOS App 开发从 0 到 1	张诏添、李凯杰
HarmonyOS 从入门到精通 40 例	戈帅
JavaScript 基础语法详解	张旭乾
华为方舟编译器之美——基于开源代码的架构分析与实现	史宁宁
Android Runtime 源码解析	史宁宁
鲲鹏架构入门与实战	张磊
鲲鹏开发套件应用快速入门	张磊
华为 HCIA 路由与交换技术实战	江礼教
openEuler 操作系统管理入门	陈争艳、刘安战、贾玉祥 等
恶意代码逆向分析基础详解	刘晓阳
深度探索 Go 语言——对象模型与 runtime 的原理、特性及应用	封幼林
深入理解 Go 语言	刘丹冰
深度探索 Flutter——企业应用开发实战	赵龙
Flutter 组件精讲与实战	赵龙
Flutter 组件详解与实战	[加]王浩然(Bradley Wang)
Flutter 跨平台移动开发实战	董运成
Dart 语言实战——基于 Flutter 框架的程序开发(第 2 版)	亢少军
Dart 语言实战——基于 Angular 框架的 Web 开发	刘仕文
IntelliJ IDEA 软件开发与应用	乔国辉
Vue＋Spring Boot 前后端分离开发实战	贾志杰
Vue.js 快速入门与深入实战	杨世文
Vue.js 企业开发实战	千锋教育高教产品研发部
Python 从入门到全栈开发	钱超
Python 全栈开发——基础入门	夏正东
Python 全栈开发——高阶编程	夏正东
Python 全栈开发——数据分析	夏正东
Python 游戏编程项目开发实战	李志远
Python 人工智能——原理、实践及应用	杨博雄 主编，于营、肖衡、潘玉霞、高华玲、梁志勇 副主编
Python 深度学习	王志立
Python 预测分析与机器学习	王沁晨
Python 异步编程实战——基于 AIO 的全栈开发技术	陈少佳
Python 数据分析实战——从 Excel 轻松入门 Pandas	曾贤志

图 书 推 荐

书　　名	作　　者
Python 概率统计	李爽
Python 数据分析从 0 到 1	邓立文、俞心宇、牛瑶
FFmpeg 入门详解——音视频原理及应用	梅会东
FFmpeg 入门详解——SDK 二次开发与直播美颜原理及应用	梅会东
FFmpeg 入门详解——流媒体直播原理及应用	梅会东
FFmpeg 入门详解——命令行与音视频特效原理及应用	梅会东
Python Web 数据分析可视化——基于 Django 框架的开发实战	韩伟、赵盼
Python 玩转数学问题——轻松学习 NumPy、SciPy 和 Matplotlib	张骞
Pandas 通关实战	黄福星
深入浅出 Power Query M 语言	黄福星
深入浅出 DAX——Excel Power Pivot 和 Power BI 高效数据分析	黄福星
云原生开发实践	高尚衡
云计算管理配置与实战	杨昌家
虚拟化 KVM 极速入门	陈涛
虚拟化 KVM 进阶实践	陈涛
边缘计算	方娟、陆帅冰
物联网——嵌入式开发实战	连志安
动手学推荐系统——基于 PyTorch 的算法实现（微课视频版）	於方仁
人工智能算法——原理、技巧及应用	韩龙、张娜、汝洪芳
跟我一起学机器学习	王成、黄晓辉
深度强化学习理论与实践	龙强、章胜
自然语言处理——原理、方法与应用	王志立、雷鹏斌、吴宇凡
TensorFlow 计算机视觉原理与实战	欧阳鹏程、任浩然
计算机视觉——基于 OpenCV 与 TensorFlow 的深度学习方法	余海林、翟中华
深度学习——理论、方法与 PyTorch 实践	翟中华、孟翔宇
HuggingFace 自然语言处理详解——基于 BERT 中文模型的任务实战	李福林
AR Foundation 增强现实开发实战（ARKit 版）	汪祥春
AR Foundation 增强现实开发实战（ARCore 版）	汪祥春
ARKit 原生开发入门精粹——RealityKit + Swift + SwiftUI	汪祥春
HoloLens 2 开发入门精要——基于 Unity 和 MRTK	汪祥春
巧学易用单片机——从零基础入门到项目实战	王良升
Altium Designer 20 PCB 设计实战（视频微课版）	白军杰
Cadence 高速 PCB 设计——基于手机高阶板的案例分析与实现	李卫国、张彬、林超文
Octave 程序设计	于红博
ANSYS 19.0 实例详解	李大勇、周宝
ANSYS Workbench 结构有限元分析详解	汤晖
AutoCAD 2022 快速入门、进阶与精通	邵为龙
SolidWorks 2021 快速入门与深入实战	邵为龙
UG NX 1926 快速入门与深入实战	邵为龙
Autodesk Inventor 2022 快速入门与深入实战（微课视频版）	邵为龙
全栈 UI 自动化测试实战	胡胜强、单镜石、李睿
pytest 框架与自动化测试应用	房荔枝、梁丽丽